Francis Keith Robins, a philosophical mathematician from Merseyside, Northwest England, possesses a singular mind from which has emerged a radical solution to the existential threats facing the world today – changing the way we think and reason. His mathematical models of set thinking and the 15 classes of knowledge have provided a framework on which a new, more egalitarian, society can be created.

A mathematical genius, Robins attributes his breakthroughs in the field of God mathematics to his hypersensitive and hyperactive mind and a life lived through the lens of mathematical principles. He has a maths degree from Bradford University and spent three decades as an auditor for the British civil service before the uniqueness of his thinking strategies were bought to light. This discovery changed his life, leading him to embark on a journey to spread the word about set thinking.

My parents, Mr Frank and Mary Robins.

Francis Keith Robins

The Fourth Coming

How God Mathematics Can Put the World to Rights

Austin Macauley Publishers™
London • Cambridge • New York • Sharjah

The story, experiences, and words are the author's alone.

A CIP catalogue record for this title is available from the British Library.

ISBN 9781035861873 (Paperback)
ISBN 9781035861880 (Hardback)
ISBN 9781035861903 (ePub e-book)
ISBN 9781035861897 (Audiobook)

www.austinmacauley.com

First Published 2024
Austin Macauley Publishers Ltd®
1 Canada Square
Canary Wharf
London
E14 5AA

To writer, Lucy Priori

www.lucyprioli.com

Table of Contents

Introduction

I see the world differently than most people. I see patterns in the chaos. I see the connections between everything, from the most mundane object to the most complicated thing in our universe – us. You trust your eyes, your senses, and your emotions to tell you about the world; I trust my mind.

This has set me apart. It has led some to tell me I am divergent, wrong even. But I think that to see what we are truly capable of, we must look beyond our assumptions. New ideas are spawned by people who do not walk the same tired, outmoded path as everybody else. People who reason in new and unique ways, and do not try to stand on the shoulders of giants.

At the core of the connections I make, lies mathematics. My ability to view the world through mathematical processes, defining my experience into sets of connected items that share characteristics (set thinking), has given me clarity into the difficulties facing us in the 21st century. It has become clear to me that only by adopting a mathematical approach and eschewing the institutions, and thinking, which are so deeply rooted in subjective decision-making, can we hope to resolve the wars, disasters, and dangerous ideologies that plague our modern world. By adopting a like-minded, mathematical way of thinking, we can eradicate the causes of devastating conflicts such as the Ukraine-Russian war. No longer divided by religious, cultural, and political differences, we can share in the rewards of an objective approach to thinking and organising our societies. By overcoming our need to understand reality, we can create a healthier inner world for ourselves, free from the anxiety and self-doubt which burdens the mind, damages our mental health, and creates the conditions for cognitive deficits and disorders like dementia.

We are all in search of a better life. Unsatisfied, we yearn for more than we have – more money, more love, more success, more purpose - convinced a happier existence is possible, if only we could find the right ingredients. Our greatest philosophers have shared our obsession.

'How should we live?' remains one of the central questions in philosophy to this day. In his Little Book of Philosophy, Comte-Sponville examined philosophical thought on everything from love and morality, to politics, knowledge, freedom, and God in search of an answer to this age-old question (Comte-Sponville, 2004). While for eminent philosopher Immanuel Kant, considering the consequences of our decisions was the basis of a life well-lived – or as he would call it 'good will'. Set thinking finally offers us the solution we have all been seeking. To the question, how should we live? My answer is simple – mathematically. We must relearn how to live according to mathematical principles, embracing the mathematical natures we were born with. Only by using our brains in this way, as God always intended us to, can we make the most of the life He has gifted us with. God's mathematics is the key to shaping the future we want for ourselves, and our societies. By using mathematical principles to think and organise our societies, as God always intended, we can transform our world for the better. This is the fourth coming of our Lord, to spread word of God mathematics and put the world to rights.

"In the first coming, he was seen on earth, dwelling among men; he himself testifies that they saw him and hated him. In the final coming [second coming], all flesh will see the salvation of our God, and they will look on him whom they pierced. The intermediate coming [third coming] is a hidden one; in it only the elect see the Lord within their own selves, and they are saved."

St Bernard of Clairvaux.

If ever there was a time to open our minds to these fresh ideas, it is now. Our world faces immense challenges. As I write, deadly wars are being fought in 32 countries, 1.8 billion people are living below the poverty line, and scientists agree our planet is dying from global warming, deforestation, and pollution. Our governments and institutions have failed to find workable solutions. Entrenched in their feuds and ideologies, they squabble while Rome burns. In the West, we cling to the idea that these are the problems of other, less developed parts of the world. We are wrong. They are creeping ever closer to our daily lives, and we have waited too long to make the changes that need to be made. Not just to our societies, but to ourselves. We see the truth of this in the daily news. Our institutions are crumbling under the pressure, our economies are stagnating, and war has broken out in Europe for the first time in decades. We see it in our families too. Half of all American adults will experience a mental illness in their lifetime, and it is only going to get worse. Our children are under pressures we never dreamt of at their age, sparking a mental health crisis amongst our young that has made suicide the most common cause of death for people under 35. Social media and celebrity culture have idealised the perfect person, and the perfect life, and our children are struggling to live up to it. A change needs to be made, and soon; And not the tried, tested, and failed changes of the past. Electing new leaders will not spark the revolutionary new approach we need. However well-intentioned this new generation of politicians may be, they are made of the same mould, influenced by the same assumptions. To enact real change, we must change the way we think, and the way we organise our societies; making use of mathematical principles and models to transform ourselves, and our world.

"It is quite likely that the 21st century will reveal even more wonderful insights than those that we have been blessed with in the 20th but for this to happen, we shall need powerful new ideas, which will take us towards significantly different directions from those currently being pursued. Perhaps what we mainly need is some subtle change in perspective – something that we all have missed."

Sir Roger Penrose.

Sir Roger Penrose is correct in the conclusion of his book, The Road to Reality. We need powerful new ideas by which to shape our future (Penrose 2005). We must harness the power of mathematics through the adoption of set thinking (Lesson 2, p18) and mathematical models (Lesson 3, p29) to capture all knowledge for any system of experience. The subtle change in perspective that Penrose calls for, that indefinable something we have all missed, is a shift in the way we think. A change in focus, from attempting to seek reality directly through the senses to concentrating on living our lives unburdened and 'in the moment', (Lesson 1, p13).

I appreciate the extraordinary benefits of this new approach because I have experienced it first-hand. My naturally hypersensitive and hyperactive mind meant I was the first of us to take this journey into a life shaped by mathematical thinking. Mathematics is based on the definition of a set (a group or list of objects that share the same characteristics and are therefore connected) and every subject can be represented by a set of knowledge, whether it be a table, a sunset, or a human being. This is how I naturally think, and so while my peers were struggling through their adolescents with the typical amount of bad behaviour and poor decisions, my mother noted that I seemed curiously immune from teenage rebelliousness. While they fought to keep up with the overwhelming flow of information and knowledge that the world threw at them, I was able to rise above it, letting information flow over my head, childlike, and get on with living my life. It is my hope that by sharing my insights others can learn to do the same.

In this book, I will take you through the process of mastering an objective thinking strategy step by step. I will offer tools to help you transform the way you think. Explain how we can teach these most critical skills to our children. Argue for the benefits of integrating mathematical thinking into our education and healthcare systems. And demonstrate how by organising our world in a mathematical way we can create a more effective, inclusive society. Through this, we find solutions to the worst existential threats facing our world today, from the destructive force of wars and terrorism to climate change, racism, and viral outbreaks.

What is God Mathematics?

Summary

God is a mathematical being. In the beginning, He created a mathematical world and populated it with creatures who can cogitate with mathematical precision. That is not a coincidence; it is part of His plan for us. A plan, moreover, which we consistently ignore in favour of subjectivity and unsubstantiated generalisations. Our refusal to use our brains in the way He intended is why, crucially, God introduced religion. The path towards a more equal, inclusive, and peaceful society lies in teaching ourselves to embrace our mathematical natures.

Legendary mathematician James Sylvester describes mathematics as the 'music of reason', and it is. Beautiful, elegant, and empowering. It is the basis of logical thought, and the building blocks of a more rational world. It is the cornerstone on which my philosophy of how people should think, and society is organised, is based. And it is part of our essential make-up as human beings.

In his wisdom, God created a mathematical world and populated it with creatures designed to think and reason with mathematical precision – us. The brain is a machine, designed to work objectively. It calculates, reasons, and draws conclusions from evidence.

Our children are born as mathematical creatures. We begin to lose this ability as we grow up and our curiosity about the world overrides our childlike instincts, but it need not be like this. As their guardians, it is our responsibility to teach children to embrace their mathematical natures so this can guide and shape their adult lives. We must teach ourselves to overcome the limits of our upbringing, shrug off the ill-fitting coat of subjectivity, and return our minds to the quieter, mathematically organised place He always intended it to be.

But first, I must dispel a myth. One so prevalent in our society today that it is holding people back from even attempting to learn to think in mathematical terms. The idea that maths is number crunching and mind-numbing equations, the domain of statisticians and accountants. That it is something to be wary of, complicated and difficult to learn. This is a fallacy. We use mathematics every day, in almost every aspect of our daily lives, without even realising it. Every time we reason, we are using mathematics.

Mathematics is the cornerstone on which my philosophy of how people should think, and society is organised, is based. Yet understanding it could not be simpler. All that is necessary to learn an objective thinking strategy is an understanding of the definition of a set - a fundamental mathematical process that we use every time we cataloguc our library books or display our groceries in the supermarket – which I will explain in more detail in later chapters (Lesson 2, p18).

"Mathematics is not about numbers, equations, computations, or algorithms. It is about understanding."

– William Paul Thurston

The reality is that maths is everywhere, and in everything. It is in the way galaxies spiral, rivers bend, and seashells curve. We find it in the periodic table, the fundamental laws of nature like gravity and electromagnetism, the laws of physics (thermodynamics) like temperature, energy, and motion, and even in our own bodies. The natural world is literally built with mathematics. Fractal patterns are the building blocks of everything from broccoli florets to pine cones, and the mathematical constant of Pi appears in the disk of the sun, the spiral of our own DNA double helix, and the pupils of our eyes. Why should we fear something that is so fundamental to who we are and how we exist?

The instinctive fear of mathematics that we have unwittingly encouraged within many of our children stems from this misunderstanding. Faced with a chalkboard of squiggly lines equations, geometry, and unwieldy calculations it is no wonder our children fail to understand how mathematics is relevant to their lives. When will they use it, they ask. If calculators can calculate faster and better than their own brains, what is the point of learning maths at all? Set thinking brings mathematics out of the classroom and into the real world, making it, at long last, relevant.

As a spiritual man, I see God in mathematics. If the world is built on mathematical principles then surely that is by His design. If our brains are rational, objective machines and we are made in his image, then surely God is a mathematical being too.

Mathematics is the language through which God speaks to us, sharing His wisdom, and a lens through which he always intended us to view the world. He would want us to think and organise our societies using mathematical principles. He gave us independence, the free will to make mistakes and create our own thinking strategies, and we chose badly, allowing our natural curiosity to subvert the natural objectivity of our minds. It does not make sense that the wonderful machine He created, the brain, should operate subjectively.

Our refusal to use our brains in the way He intended is why God introduced religion, to fill the void in our own understanding of His mathematical creation and provide the structure that our ignorance has cost us. It is unfortunate that our human failings have led to religion becoming the cause of so many of our conflicts. The adoption of objective thinking strategies will do away with the need for such religious ideology. Faith in God together with praise for His Prophets will, naturally, remain, but the ideological differences between those of faith, which are so frequently the cause of conflict, will not.

I do not ask that you share my faith, of course, only that you do not disregard my mathematical logic for operating an objective thinking strategy. But it makes little sense to me that God would give us the tools to think and reason objectively if He did not want us to use them to shape our thinking and our world. Especially when the benefits of this approach are so great – a more content, peaceful, and inclusive existence.

The Christian, Judaic and Islamic traditions all teach us of the danger of seeking reality. In the story of Adam and Eve, the First Couple are banished from the Garden of Eden for eating from the tree of knowledge. Their sin was to attempt to seek reality directly from the senses, losing touch with the magical mathematical world. Yet we have ignored the warnings inherent in these religious teachings, with devastating consequences.

By changing how we think we can change the world, and finally live up to the potential He gifted us with.

A New Way of Thinking

Summary

Mathematical thinking offers a solution to the myriad problems facing the modern world. Subjectivity has proved damaging to us, our children, and our societies. A new way of thinking must be adopted globally if we are to create healthier, happier lives for ourselves and the better world we deserve.

Subjective thinking is the root cause of many problems facing our modern world today, from racism and inequality to ideological warfare, the mental health crisis, and teenage delinquency. Tackling these issues will require a new way of thinking to be adopted on a global scale, integrated into our institutions and societal structures, embedded within our healthcare models for the prevention and treatment of mental illnesses, taught to our children, and embraced by our leaders so they can implement the changes our world desperately needs.

The consequences of subjective thinking for our children are especially worrisome. One in six children in the UK are now believed to have a mental health condition, and last year 13,800 of them were cautioned or sentenced for criminal activity, including 3,500 knife offences. The recent youth riots in France, so reminiscent of the London riots in 2011, are yet another example of the undue pressure being placed upon today's young people by our failure to teach them mathematical thinking.

Without proper instruction in how to think, our children have been left rudderless and feeling abandoned. Their desperate and futile attempt to seek reality has left them overwhelmed by the barrage of information the modern world exposes them to. In their confusion, and with their minds cluttered by untruths and misrepresentations, they lash out and are drawn into antisocial behaviours. The subjective thinking strategies we have encouraged through example have left them carrying the weight of the world on their shoulders, crushed beneath the perceptions they create when they ought to be focused on living their own lives. Consequently, their mental health is suffering, and juvenile crime, delinquency, and other violent behaviours are on the rise.

It is not only our children's mental health that is suffering. Our subjective natures have led us all down a dangerous path. By seeking to understand everything, we have become overloaded with information and without any filters for what is real and what is useful. Without rational tools to guide us, we make rash judgements, assumptions, and mistakes. We place too much emphasis on things that do not matter, like the way a person looks, or the perceptions people have of us, and not enough emphasis on the things that do. Perceptions are at the root of the prejudices and inequalities that beset our world. Racism, sexism, and gender biases all arise from a misrepresentation of reality based on faulty assumptions and incomplete facts made by people who have attempted to understand the world and failed to realise that they do not.

Mathematical thinking can empower us. With our minds unburdened by unnecessary data, we can focus on what really matters and lead healthier, happier lives. Over the next three lessons, I will endeavour to teach you how to adopt this new way of thinking and incorporate it into your daily experience.

Lesson One: Do not Seek Reality

Summary

We are the creators of our own false reality. We must learn not to create perceptions of the outside world in our minds, to live in the moment, and allow information to flow over our heads, childlike which means you are not seeking to escape reality. The first step towards developing an objective thinking strategy is to accept the fallible and unreliable nature of the so-called reality which we create from information gleaned through the senses, and instead, concentrate on making the most of our lives.

The first step towards adopting an objective thinking strategy is to teach yourself not to seek reality from information received directly through the senses, because the set of information on which it is based can have false characteristics. Acknowledge and accept that the perceptions of the world you create in your mind can be wrong and incomplete. That an accurate picture of the world is unobtainable and unnecessary. That the only reality that matters is the one immediately facing you. Let yourself live in that moment, letting information flow over your head, without seeking to look beyond.

Reality is an illusion. A construct of our own minds and perceptions. The greatest philosophers in history have been telling us this since antiquity. Modern-day quantum theorists were forced by the weight of evidence to come begrudgingly to the same conclusion. Yet we cling to the notion of reality with both hands, desperately seeking the stability and sense of permanence that reality promises. We seek the comfort of believing in what our eyes tell us, even when it is proven time and again to be false. We insist on drawing together the incomplete data, fragmentary evidence, and transitory perceptions of our experiences and labelling them as the truth, even though we know it is unreliable. Warned of AI bots disseminating fake news and media moguls with their own agendas, we nevertheless choose to believe in the information we receive when the truth is that the world is too complex a mixture of facts, opinions, and inaccuracies for us to understand it based on our experience alone.

In 830 BC, Plato posited the allegory of The Cave (Magee, 2001). He imagined a group of men sitting around a fire, watching the shadows playing across the cave walls of their prison. These men had never walked beyond the cave entrance. They had no concept of sunlight or open fields. The shadows were their only insight into reality. They gave the shadows names. Draw conclusions about the nature of the universe from the evidence the shadows presented, never knowing that their representation of the world was incomplete, and that the reality they had constructed was wrong. In a very real way, we are those men. Making assumptions about the nature of reality based on what we have been taught, what we read in the papers, and what we experience in our everyday lives. Just like them, the picture of reality we construct in our minds is based on incomplete information, falsehoods, and presumptions; and like them, we have no idea how wrong we are.

As Bryan Magee says, “Our direct experience is not reality but what is in our minds” (Magee, 2001). Our reality is dictated by our perception of it, and our perceptions are forever changeable. All it takes is a new piece of knowledge, an intriguing new idea, and our perceptions shift, transforming our view of reality in the process. For two millennia, people used the term black swan to mean something impossible, non-existent, a laughable absurdity. Coined by a Roman satirist, the phrase took its meaning from the fact that all swans were white. Then in 1687, Dutch explorers discovered black swans in Western Australia, making them a perfect metaphor for how pointless an endeavour seeking reality really is. The knowledge we have of the world is unavoidably incomplete, and therefore so are our perceptions of reality. “We have to recognize that the world is not something sculptured and finished, which we as perceivers walk through like patrons in a museum; the world is something we make through the act of perception,” says Terence McKenna.

In his book The View from Nowhere, Thomas Nagel noted that “one of the strongest philosophical motives is the desire for a comprehensive picture of objective reality” (Nagel, T, 1986). He later acknowledged that “since the idea of objective reality has to be abandoned because of quantum theory anyway, we might as well go the whole hog and admit the subjectivity of the mental.” Notable philosopher Thomas Lewton reached a similar conclusion, venting his frustration that "it can seem as if there is an insurmountable gap between our subjective experience of the world and our attempts to objectively describe it." The apparent impossibility of establishing a shared and objective picture of reality has stumped philosophers who are attempting to understand human consciousness. As renowned philosopher David Chalmers admits, "the hard problem of consciousness is to describe the seemingly intractable problem of subjective felt experience" (Chalmers, 2022). 25 years ago, he famously bet esteemed neuroscientist Christopher Koch that all attempts to identify the neural correlates of consciousness would fail, and that by 2023 scientists would still be unable to explain the brain mechanisms which underlie consciousness (Chalmers & Koch, 2023). Recently, Koch stood before an audience of his peers at the 26th annual meeting of the Association for the Scientific Study of Consciousness at New York University to admit that Chalmers was correct. Prematurely, as it turns out. Unbeknownst to them, when they made their bet in 1998, I had already solved the mystery of consciousness – more than three decades earlier [p16]. Still in my teens, I had no idea at the time that by not seeking reality which can be represented as a set of knowledge which can have false characteristics and allowing information gleaned through the senses to go over my head I had discovered the connection between reality, consciousness, and mathematics (Penrose, 2006), and treating the experience which surrounds us as a set of knowledge, or that I had already proven wrong Chalmers' assertion that objective thinking is impossible by not creating perceptions. My discoveries showed that we can establish an objective picture of reality, by using the mathematical models outlined in this book (p31). The issue isn't whether we can establish reality directly through the senses, but whether we ought to.

Philosophy poses two fundamental questions: What can we know? And how can we know it? (Thompson, 2003). I would argue these are the wrong questions. Of far greater relevance to our lives and happiness is what do we need to know? (p49 p51). Understanding reality is, for the most part, entirely unnecessary. Aside from a few specific areas of our lives (which I will address in p49 p50), we can live perfectly happily and successfully without attempting it. Yet curiosity remains our downfall. Discontented with the narrow reality we experience in our day-to-day lives, we strive for greater knowledge, processing data with unrelenting speed.

An objective thinking strategy makes the search for reality as unnecessary as it is unproductive. Free from the need to worry about the outside world, you can concentrate on living your life, and use the thinking tools outlined later in this book to gather only those elements of reality that serve a practical purpose (p20, p31).

Rather than seeking reality, which is at best unfathomable, and at worst, not there to find, you can learn to unclutter your mind of the generalisations and false perceptions that so often lead you wrong. **By adopting this childlike approach as a youngster of allowing such information to go over my head, the result of change over to adulthood was my consciousness and subconsciousness came in line.**

Instead of struggling to shut out the barrage of data that batters your senses day in and day out, you can train yourself not to seek it in the first place; to shun the subjective interpretations of the world that are overwhelming you, and think objectively, using mathematical principles.

You can learn to shelve your curiosity. To take from the past only that which will help you learn from past mistakes and the rules you need to make the most of your life. To think of the future only as much as is necessary to make plans. And by taking one experience at a time without referring backward or colouring your experiences with an inaccurate haze of learnt preconceptions, your thinking can remain untainted by bad memories, past experiences, and history. This does not mean that our society should forget its misdeeds, or the historic events that shaped it. Rather, such memories are unnecessary, and unhelpful, for us on a day-to-day basis.

With these objective thinking strategies ingrained, our interactions with the world will alter. Asked whether we trust the police, or what we think of the current state of the NHS, as an objective thinker we will answer that we do not have a viewpoint. How could we? And crucially, why would we need to do so? After all, without a complete audit, we lack all the information at hand to make a comprehensive analysis. We know only that which we have received through our senses, some, or all, of which may be wrong. And if we have no immediate need of a policeman or the NHS, then having an opinion of either will be an unnecessary distraction from living our lives.

This will not, however, be a simple accomplishment. The drive to seek reality has been ingrained in you, fostered and encouraged throughout your life. Our culture and education system actively encourage our constant search for reality. From childhood, we are taught to be sponges, absorbing as much information from our surroundings as possible. Social media has tapped into this urge and worsened it by a significant degree. Neuro-linguistic programming could, in time, be adapted to offer an easier path towards changing the way we think, but for now, we must face the challenges with an open mind and a willingness to try our best. The difficulty of the task ahead should not dissuade us, especially not when the future of our children and our society is at stake.

This childlike way of thinking will come far more naturally to our children than it does to us. Born as mathematical beings, they already possess everything they need to use their brains in the way He intended. We need not teach them to think differently, only to prevent them from outgrowing their own instinctive rationality. And whilst we teach them to think and live in this way, we must also endeavour to do so ourselves. We must set the example for them to follow and fully understand how we can support them in holding onto their mathematical natures.

Lesson Two: Apply Set Thinking

Summary

Set thinking can equip you to hold onto your mind's childlike flexibility, uniting the conscious and subconscious halves of your mind, exposing you to new possibilities, bringing gut feelings to the fore, and enabling you to apply reason and method to your choices. Four set thinking tools are presented to aid you in making the most of your daily life – classify and expand, ordering, consequences, and permutation and combination. With them, you can learn to live more mindfully, consider the consequences of your decisions, and become better at improvising, planning, prioritising, and creating routines.

The second thing you must learn if you are to adopt an objective thinking strategy is to begin treating the reality that surrounds you as a set of information to investigate (or not, if that is your choice). Mastering this way of thinking, which I call set thinking, will assist you in your attempt to train yourself not to seek reality and to remain present in the moment. Your mind cannot become cluttered with generalisations and perceptions if it is already occupied with the task at hand.

Your consciousness is divided. The subjective thinking strategies you have been taught since childhood mean you exist between two states: your conscious self, and your subconscious, where all your hidden agendas and practised behaviours reside. The disparity between these two sides of your mental landscape becomes obvious when you consider a simple example. When you first learn to write, you are aware of every letter you meticulously place on the page. Your attention to detail is precise, your focus intense, because you do not want to make a mistake. Over time, however, the skill becomes automatic. You can scrawl out a note to your kids or your partner and stick it to the kitchen fridge with hardly any attention at all.

This lack of focus means we miss most of our daily experiences. We live so much of our lives unconscious of what we are doing that we could conceivably go through an entire afternoon barely noticing the interactions our body is participating in or the multitude of choices we have made. We put milk in our tea because that is what we always do. We put butter on our sandwiches because we always have butter. But what if we stopped to think about what we were doing? We may like tea without milk. We may prefer a glass of milk on its own. We might even decide to try something new – orange juice, a smoothie, or a cold beer. Applying set thinking to such decisions makes you more aware of your choices and enables you to apply reason and method to them, by using the set thinking tools explained later (p20). By adopting an objective thinking strategy, your conscious and subconscious become one. Your subconscious awareness, the gut feelings, and intuitions which we all rely on so heavily, become fully integrated into the objective thinking strategies you employ. With your subconscious thoughts now co-existing within the realm of the conscious, your mind is opened to new possibilities.

What is Set Thinking?

In simple terms, a set is a group or list of objects that share the same specific characteristics and are therefore connected. Sets are the building blocks of mathematics (Spiegel, 1963). They provide the foundations for all branches of maths and can be used for collecting and classifying objects. Typically, this involves numerical sets, such as sets of even numbers, odd numbers, or primes. However, the mathematical properties inherent in language make it possible to apply sets to any object. There is a one-to-one mathematical relationship between language and reality. The word 'finger' has the same characteristics as the finger itself. Therefore, it is possible to gather all knowledge for any system or experience through mathematical means. There exists a parallel mathematical world.

Let us take the chair you are sitting in as you read this book as an example. The chair has a seat, legs, armrests, and a cushion. All these elements form a set. They are connected by being elements of the chair. Cataloguing all aspects of the chair in this way is called a downward set. Once you have identified every element of the chair (every element in your set) you can be said to have a complete set.

Your chair is also part of the room you are in. Along with all the items in that room, it forms a set of the room. We call this raising the classification. This new set could include a table, more chairs, a rug, a bookcase, the floor, ceiling, and windows. All these objects are connected and are part of a set. You can now raise the classification further, to create a set of the building you are in, including not only your room but the kitchen, bathroom, stairways, even the people inside the building. This is called an upward set and involves stretching out our imagination to envision all possible elements of the set that exist in the outside world. Raising it further still, you can create a set of the street that building is in, the town it is in, the country it is in, and finally, the planet it resides on. Want to raise it even more? This planet is part of a set including every planet in the solar system. A set can therefore represent any member of reality from the smallest element known to exist, the monad, to the entirety of human knowledge and experience. Everything can be said to be connected, and part of a set. There is no end to the number of sets you can create with a given object, and every aspect of your experience can be classified and analysed using set thinking, from the items on your dinner table to your journey to work. Free will naturally applies, however. Though you can choose to apply this thinking strategy to a task, you can also choose not to. At each step of your day, you make the choice whether to apply set thinking or not.

"Mathematics, rightly viewed, possesses not only truth but supreme beauty — a beauty cold and austere, like that of sculpture, without appeal to any part of our weaker nature, without the gorgeous trappings of painting or music, yet sublimely pure, and capable of a stern perfection such as only the greatest art can show."

Bertrand Russell

Set Thinking Tools

Now that you understand what a set is, the next step is to begin applying set thinking to your everyday life. My research has identified four mathematical mental tools that will support you in this endeavour. I will run through each in turn, providing examples of how you can use them to create the foundation of your new mathematical way of thinking.

Tool 1: Classify and Expand

By classifying, you are defining the characteristics of the set (or list of items that share the same characteristics), as we did earlier. By expanding, you are seeking members of that set. Using this thinking process increases your knowledge base mathematically and helps you to expand your horizons and see all the possibilities life offers.

You can create a set for anything. The reality we experience consists of a set of elements, for instance, the atmosphere, a seat, a person, a computer, the wall, carpet, and windows. Each element can be described as a set of knowledge. The computer could be broken down into a set including the keyboard, mouse, and screen. The person could similarly be classified as a set that contains their eye colour, hair, and personality traits. Game show contestants frequently use these classify and expand tools to widen their knowledge base and memorise large numbers of facts in the form of sets or lists. For instance, if our game show contestant is asked a question about a particular US President, he could classify and expand a list of all Presidents with relevant or connected information until he has a complete set.

Bear in mind that technological solutions can serve the same function (are connected) as non-technological ones. They form a set, for instance, sending a Christmas email instead of a Christmas card, telephoning a friend rather than visiting them, or walking instead of taking the car. You will therefore want to consider these while creating your complete set. As an example, let us classify (create a set) for your journey to work in the morning – a way of getting from A (your home) to B (your workplace). Now that you have defined the characteristics of your set, you can expand it – seek other members of your set. You could walk to work, take public transport, drive, or take a taxi. You can also refine it to create a more limited set, for instance, going to work by public transport. Now your set will include all your public transport options – taking a bus, the train, and a ferry.

Let us consider another example. You always have porridge with honey for breakfast. You could use the process of classifying and expanding to get you out of this rut. You classify a set of edible items in the house that go with porridge and expand it. Your set could include sultanas, blueberries, bananas, cherries, and spices like nutmeg and cinnamon. Now you have a variety of options to choose from.

This process of classifying and expanding can be applied to any action in your daily routine, from getting up in the morning, to making breakfast, to going to bed, and anything in between. You can use them to help you think through the actions and decisions of your life, bringing them into the conscious rather than the subconscious, considering the choices you are making and the alternatives available to you.

It is important to note, however, that it is possible to apply set thinking wrongly. If, for instance, you have a bad experience with a policeman, you could conceivably apply classify and expand processes to conclude that all policemen are inherently bad. It is crucial, therefore, that you learn to apply these tools effectively and remain vigilant to the creation of generalisations. You may realise that you cannot generalise as each person and situation is unique although they can be gathered into sets.

Tool 2: Ordering

Now that you have identified all the options available to you, you will need to choose one. This is ordering - selecting a course of action from the multitude on offer. You do this by making connections (forming sets) between the different items in the set, comparing them, and prioritising them. For instance, X = 3X = Y, where X and Y are the two alternatives. Y is three times greater than X.

You are offered a slice of cake, but which slice should you choose? You compare them. One slice has more icing. Another is larger than the rest. A third is smaller but has a cherry on top. You decide you are not very hungry, and you really like cherries, so you choose the smallest slice. You have compared the choices on offer, and prioritised the aspects that matter most.

You are playing with your child when your mobile phone rings. Do you A, answer the phone, B, ignore it and continue caring for your child, or C, attempt to do both? You compare your options and prioritise. What is more important, the call or your child? Are you expecting an important call that cannot wait? What are the feelings of the child when in general the mobile phone usage takes priority? You can also raise the classification to other situations where there is a conflict between your mobile phone usage and your family obligations, such as mealtimes, for instance. In all situations, you consider the consequences of decisions and what mistakes can be made.

These are comparatively petty examples, but the same process can be applied to scenarios with a far greater impact. You are on a committee that will decide where best to allocate government funds. Do you spend taxpayers' money on space exploration or medical research and climate change measures? You can make this decision using mathematical principles. Classify and expand your set to include all possible projects that you could fund with this money, and then order them by connecting, comparing, and prioritising.

Tool 3: Consequences

Life is a series (or set) of decisions that can be represented as an equation: decision = consequences, where both sides form a set. Considering, where possible, the consequences of your choices is an essential part of the decision-making process. It can also be useful for ordering, so much so that you may want to consider the consequences first. Although I have presented the set thinking tools in a specific order here, you are free to apply them in whichever order best suits your needs. I would argue, for instance, that politicians would be wise to consider the consequences of projects before ordering their options.

I remember as a child coming home from church one day bereft. I admitted to my mother that I could not keep to all the rules set down by the Methodist church I had been raised in and asked if I could consider the consequences of my actions instead. I did not hold much hope she would agree. To my surprise, and relief, she smiled. 'That is all God would want,' she explained. It is the consequences of our actions that give them meaning, and that determine the impact we will have on this world and yourself.

Tool 4: Permutation and Combination

This thinking strategy is helpful when you are faced with more than one choice. It enables you to make the most efficient use of your time by combining tasks, and to select multiple choices. Combination and permutation can be expressed mathematically as members of a set which all have the characteristics of achieving a common objective.

Let us take your morning routine as an example. You need to get dressed (task A) and make breakfast (task B). The most efficient use of your time is to perform your duties while waiting for another action to be completed, so you could put bread in the toaster (B) and partly dress (A) while waiting for it to cook. After buttering your toast, you turn on the kettle (B) and finish dressing (A) while it boils, then finish making breakfast (B). Or alternatively, fully dress (A) and then make breakfast (B).

You can also apply it to your porridge dilemma from earlier. You classified and expanded your choices of porridge toppings and discovered a range of options you did not know you had. Then you ordered them, comparing and prioritising your options – let us imagine you chose a banana. With permutation and combination, it is not necessary to choose only one. You could have banana and cinnamon, for example, or banana and blueberries.

This process most likely feels very familiar to you. It should. We perform permutations and combinations quite naturally throughout the course of our lives. You do it every time you eat a meal. Should you eat the vegetables first or the meat? Or have a piece of beef, then a carrot, then another piece of beef? Or perhaps put both the beef and the carrot on your fork at the same time?

Set Thinking in Action

Let us now apply everything we have learnt to real-world situations. Mathematical thinking can be used to capture all knowledge for any given situation, so you can apply set thinking to any, and all, experiences if you choose to. We routinely apply set thinking throughout the course of our lives. This is how the brain naturally operates, seeking sets, expanding, and then choosing a course of action. If you have ever made a bucket list, a wedding list, a rota, a shopping list, or used a list of ingredients to make a recipe, you have been using set thinking. Developing an objective thinking strategy merely requires that we make a conscious decision to apply set thinking to our everyday experiences.

A telephone directory is a good example of set thinking in action as it is a list (or set) of telephone numbers. A telephone book is made up of a multitude of sets. You could make a set of the telephone numbers of your friends, for instance, or of experts, people in particular professions such as doctors, dentists, or plumbers, establishments such as councils or libraries, or any other classes of your choosing. You could refine it to create more limited sets, for instance, a set of landline numbers, mobile numbers, email addresses, or postal addresses. You could connect with a library for directories or sets of telephone numbers, or with the internet as a source of telephone knowledge. You could also analyse the entries mathematically breaking down the words and numbers into sets of letters and numbers which can have various sets of properties – their colour, size, shape – and expand these out to include the range of all possible options. The colour of the entry may have some significance, or it may not – or to express this mathematically, the colour characteristics of the set may have meaning or not. With so many letters and numbers on offer, you could also apply combination and permutation and ordering to your sets.

I use set thinking every day. It is normal to me, as I hope, one day, it will become normal to every man, woman, and child in our society. As such, it has become the lens through which I make sense of, and interact with, the world around me. The events I experience throughout the course of my day, even the way I choose what to have for breakfast or what clothes to lay out for the next day, are all considered using mathematical principles. My daily routine therefore offers a preponderance of examples of set thinking in action that I could share with you. From the moment I rise from my bed to start my day I am faced with an opportunity to apply set thinking to my experiences, by using classify and expand tools to create a complete set of my morning ritual including washing my hands, making breakfast, and brushing my teeth. As you begin to adopt mathematical thinking you too will find that opportunities for applying objective thinking are all around you, in every sight that attracts your attention, and every experience you live. In experiences both big and small, monumental, and mundane.

While on the train yesterday, I saw a message board warning travellers to please mind the steep step down to the platform. Naturally, I classified and expanded, defining the characteristics of the set and then seeking other members of the set. I asked myself, does the message have truth characteristics? Is there indeed a steep step down to the platform? I considered the overall set, selecting the individual message and then expanding to other potential messages, and the sets within, such as the words, numbers, or other elements that the message contained.

By classifying the message as a warning of a steep gap to a platform I was able to conclude that the same message should be given for other stations with steep steps on my journey, and raised the classification to include all stations nationally with similar conditions until I had a complete set. I also connected the reason for the message was health and safety legislation, thereby expanding to include other relevant messages with the same cause such as warnings to take care on the escalator. Alternatively, by classifying it as a message on an information board I could expand to create a set of other classes of messages on boards, including those alerting passengers to an upcoming tourist feature, what the next station is, where they can connect with other rail services, what to do in the event of an emergency onboard, or where the train terminates. I could also raise the classification to note that the notice board was part of the carriage and expand to create a complete set including the seats, information boards, and people. I could then consider the consequences of the messages I had identified. Were they necessary? And what might the impact be if I use my free will to ignore their advice. I could connect with the relevant classes of people who might read these messages, including passengers with hearing difficulties or sight impairments and foreign travellers, considering what actions might be necessary to make the messages accessible to their needs, such as utilising verbal messages as well as message boards or printing them in multiple languages.

When seeking internal sets (sets within) I could note that the written messages are a set of words connected with a meaning singularly and overall. Each word is also made up of letters which are made up of language, size, colour, and shape characteristics. These could be expanded to create a variety of sets (different colours, sizes, shapes, and languages) so I could select the appropriate one from the available possibilities.

Another opportunity to apply set thinking occurred to me as I was walking down my High Street the other day. My local council is fond of creating floral displays along our streets, and it occurred to me they could use set thinking to design more original and interesting displays. Firstly, they would define the characteristics of the plant holder, classifying it as a simple shape for containing flowers, and expand it to include all manner of available shapes (circular, square, rectangle, hexagonal etc.) until they had a complete set. They could then order and compare these options to decide which is best, considering the variable consequences such as how they look, which shapes are available, their respective costs, and how fit for purpose or practical they are. They could also apply permutations and combinations, perhaps using some circular-shaped plant holders and some rectangular ones. The pattern of the flowers could also connect (or form a set) with an event, such as using red, white, and blue colours to represent a celebration of the UK's success. Considering the internal sets could also help identify and select which materials to use (wood, plastic, ceramic, or metal), what colours, what size (considering such consequences as whether they would restrict access by mobility vehicles or present a health and safety risk to pedestrians), which plants to use (annuals or perennials, seeds or grown plants, scented or unscented varieties), and what patterns to plant the flowers in.

While watching a BBC2 program about lobster harvesting in Canada I yet again instinctively began creating a set for the relevant characteristics for breeding lobsters in tanks, considering the characteristics the water must hold to release them into the wild, and the areas of water that might match them; evaluating the possible mistakes that could be made such as selecting the wrong types of lobster, the likelihood of them occurring, and the precautions that might be taken such as assigning responsibility for the decision; considering which rules apply and whether they ought to be obeyed, and the consequences (such as the impact on wildlife) if they were not; assessing the finance elements such as expenditure, income, and profitability; and considering which sources were available to find the information I needed.

You can use the same thought processes for everything in your daily experience. Your movements, for instance, involve a set of parameters including speed, type, and direction. You can classify and expand your speed from standing still to your top speed, the type of movement can be represented by a set including hopping, walking, and running, and a complete set of direction could include left, right, forwards, backward, sideways, and angles from 0 to 360 degrees. You can also apply permutation and combination. You could hop forwards and then run sideways, for instance. The brain does this naturally. Hitting a snooker ball is another example. Again, you have a set of parameters which now include the weight of the shot, the direction of the cue, the angle at which you hit the ball, and the properties of the ball itself. Or if football is more your type of sport, you can identify a set of parameters for kicking a football, such as the weight of the shot, the direction of the kick, the angle the boot strikes the ball, the part of the boot used for the kick, and the properties of the ball and your footwear.

Set thinking becomes particularly impactful when we apply it to people. By classifying and expanding a person, or group of people, we soon discover that while everyone is unique, they are at the same time based on the same template. They form a set. They are connected. By using an objective thinking strategy, we develop an awareness of the ways in which we are, all of us, alike. Without generalisations and stereotypes to cloud our judgements, we can see people as they really are. This is key to the prevention of stereotypical thinking and discriminatory viewpoints (p78).

Adopting mathematical principles can also help us connect with every facet of an experience. Our mathematical nature can inspire us, challenging us to do more and be better. Even a walk can become an opportunity to better understand our own nature and make the most of life (Lesson 1, p13).

You see a berry bush, and by expanding a set, can connect with wine making and jam. You make the connection that your smartphone can also be used to take pictures and decide to take up photography. The view could inspire (or connect) you to paint a picture or write a poem. You could connect with fellow walkers in conversation, pointing out beautiful scenery, interesting features, or sharing philosophies. You could evaluate your own fitness levels and consider new health regimens. Or you could take the opportunity this moment of reflection offers to review your own patterns of behaviour and how your parents' behaviours have shaped yours, providing insights into your own identity which could ultimately enable you to design experiences that suit you, rather than copying the choices of others – which you should only do if the person you are imitating can be classed as an expert in the task.

You could also use a mathematical thinking strategy to consider alternative routes you could travel. Walking across a field, you could ask yourself do I have a complete set of possible exit routes? Perhaps you could try opening the gate rather than going over the stile? Driving home from your day out you could consider, are there other members of your set of routes from A to B? You might also want to consider the consequences of your chosen, or alternative, routes, such as the time taken, petrol consumption, or the opportunity to drive through an area you have never seen before, making new discoveries.

Artistic pursuits, such as painting, sculpture, and poetry can be used to expand our thinking by helping us make connections. We see a homeless man on the street and are inspired or connect to paint a picture of a recumbent figure on a bench or to write a poem on the experience. Similarly, our senses – sound, touch, taste, smell – can encourage connections. By adding a touch of perfume to Christmas cards, we can evoke memories of childhood festivities, while the sound of a kettle boiling can tell us that the water is ready to pour into the cup. The possible applications for set thinking are endless.

You could, for instance, apply the four set thinking tools to checking a piece of work. Imagine you find a mistake (X). You can then classify and expand to seek the same type of mistakes (X) in the rest of your work. When you have identified all the similar mistakes, you have a complete set. You can then scour the work for any other mistakes, and if you find one (let's call it mistake Y), you repeat the process and continue to do so until all the mistakes in the document have been found. Free will, of course, means you can choose to stop searching for mistakes at any time. You can also apply it to our tea with milk routine from earlier. Rather than putting milk in our tea simply because that is what we always do, we can instead classify cold drinks we have in the house and expand to create a set including orange juice, a smoothie, and a cold beer. We can apply permutation and combination to give us a complete set of possibilities – tea with milk, milk on its own, tea first followed by milk, and milk followed by tea. We can utilise improvisation as needed, for instance, by using the clean end of a dirty spoon (X) to stir our tea instead of a clean spoon (Y) where X and Y form a set, despite X not being designed for the Y function. And we can use ordering to prioritise and select our perfect drink, accounting for the consequences of our choices, for instance, will using the milk mean we have none left for tomorrow's cereal?

Our leaders can also use set thinking to help make vital decisions that will shape our society into the future. Ought they invest taxpayers' money on space exploration, transport infrastructure, preventing climate change, or medical research? What is of greater consequence, immediate benefits, or long-term gains? Public opinion, or project success? Every decision our government makes can be considered using set thinking, creating complete sets of all the possible actions, ordering them in terms of importance, and evaluating their consequences. In later chapters, I will demonstrate how reorganising our societies along mathematical lines can transform our world for the better (p66), and the catastrophic impact that not doing so has had in the past (p61). But for now, let us return to your third and final lesson in how to develop an objective thinking strategy of your own.

Lesson Three:
When to Seek Reality, And How

Summary

There are times when you need to determine reality. Mathematical tools which help us identify when to seek reality, and how to do so effectively, are therefore critical. The 15 Classes of Knowledge and set thinking provide a mathematical means of gathering all knowledge about any experience or system. They equip us to handle the complexities of the world without falling prey to the subjective interpretations that set thinking is designed to prevent.

Once we have grasped the concepts of set thinking and are comfortable using an objective thinking strategy, it is time to begin considering when, and how, to seek reality.

Reality may be illusionary, but there are times when it is impossible to avoid, when our work or responsibilities require us to look beyond ourselves and our immediate reality to view the outside world as it really is. For most of our day-to-day lives seeking reality is nothing but an unhelpful distraction which prevents us from living fully conscious of the moment, but there are times when more strategic thinking is required. The most critical issues facing the world today cannot be addressed effectively without a rational review of the facts. How can we create an affordable NHS that cares for the health needs of an entire population, for example, or shape an international strategy for tackling climate change? It is therefore crucial that we learn to differentiate between those instances when knowledge can offer valuable and useful insights and those which are driven by mere curiosity. For this, we will need a method for systematically establishing reality which is not reliant on inherently faulty information derived from the senses – the 15 classes of knowledge.

"Mathematics allows for no hypocrisy and no vagueness."
Stendhal

Reality Seeking Tools & Mathematical Models

The 15 classes of knowledge constitute a mathematical model to gather all connected knowledge for a system or experience. As establishing truthfulness requires the application of set thinking – life is, after all, a series (or set) of experiences which can be represented by a set of knowledge - they are designed to be used alongside set thinking to equip you with the tools you need to objectively analyse any experience, system, or situation using mathematical means. They include a class of knowledge specifically designed for establishing that the system or experience in question has truth characteristics.

As with set thinking, the mathematical properties are based on the natural symbolic mathematical properties inherent in language. Therefore, the connected classes of knowledge are identified together with their associated or connected words. The 15 classes of knowledge can be used as a framework (or blueprint) to capture all the required classes of language associated, or connected, with them to create a system. Or to put it another way, to gather all associated words for a given experience or system. By studying patterns in language representing these experiences my research has highlighted six classes of knowledge that are of utmost importance when determining reality – Reality, Rules, Action, Time, Place, and Variable Consequences. I call these the primary classes of knowledge. An additional nine secondary classes of knowledge make these thinking tools comprehensive enough to be useful analysing any situation we may find ourselves in.

In this lesson, I will demonstrate how you can use the 15 classes of knowledge to gather all the connected knowledge you need to make effective, unbiased decisions. To illustrate its real-world application, I will apply the mathematical model to a social policy question that is becoming increasingly relevant in today's war-torn world – the resettlement of refugees.

Imagine you have been tasked with resettling Ukrainian refugees in the UK. In our subjectively driven society, any response to this issue will likely become highly charged and open to debate and speculation. Strong opinions will be held on either side of the debate. Politicians may be steered more by public opinion than by the facts, and the voters' influence by media sensationalism and a lack of accurate information. The issue can become clouded, rash judgements made, evidence ignored or misrepresented, and weak, ineffective solutions chosen. As you will see, the 15 classes of knowledge and set thinking offer a better way.

The Primary Classes of Knowledge

Reality

Firstly, identify the subjects of your set. In other words, who and what? This can include all humans, animals, vegetables, inanimate objects, vegetation, fungi, fire, atmosphere, gravity and weather.

Now that you have your subject, create the appropriate subsets. This involves separating your subjects into related categories, for instance people with (and/or without) computers, people with mental or physical disabilities, or people who do not speak English. The subsets you choose will of course vary depending on your subjects and the situation, but as the human condition offers great variety, the more subsets you can identify the more accurate your understanding of reality will be. This process can be expressed mathematically as asking yourself: have I a complete set of relevant (or connected) categories of people? Doing this will make your project more inclusive and help prevent a one size fits all approach by accounting for all the vagaries of the situation and people involved. Be sure to consider all the data and identify knowledge associated with each category for use where necessary. Depending on your objective, these could include your subjects' personal requirements (such as food and shelter), and their personal characteristics (such as age, gender, and skills). There will also be possible connections with other people – friends, family, and neighbours. Listing the associated knowledge can help you to make better decisions and create more effective models. Use the set thinking tools we learnt earlier for this. Classify and expand to make the most of your analysis. Permutations and combinations can be applied here too. Bear in mind your subjects will almost certainly fit into more than one subset.

Now consider any other subjects. Consider the classes of connected organisations, such as charities, local businesses, and the like. It is important to account for all the situations that could potentially arise for your subjects to create a comprehensive set for this, for instance they may become ill, change occupations, or become unemployed. Ask yourself at each stage the likelihood (probability) of each subset occurring.

Example: Ukrainian refugees

Let us put this into practice by using our Ukrainian refugee example. Your subjects will include the refugees themselves, their relatives, the people who invite them into their homes, and members of the wider community who may be affected, as well as the doctors, dentists, teachers, and social workers who will be providing them with the services they need. Be sure to consider all their personal requirements (such as toilet facilities, exercise, shelter, clothes, food, toys, and medicines) and personal characteristics.

You will need to create several subsets to account for all the variety your refugees present. These could include subsets for language – categorising refugees who speak English, only Ukrainian, or multiple languages. You could create more subsets for their occupations, education level, the number of children they have, health conditions, age, and gender. Using your four thinking tools you can then classify and expand upon all these subsets. For instance, people with health conditions can be expanded out to list all possible ailments your refugees have (or may develop), and again to create subsets of those in need of regular medication and not. You can then utilise permutations and combinations to identify more possibilities and subjects who fit into multiple subclasses. By classifying and expanding other living quarters you could develop original ideas for rehousing your refugees, for instance by classifying living quarters with X characteristics and expanding to using tents, barges, or barracks. By then considering the consequences, and ordering these ideas, you will separate those that are unfeasible with those that can be made to work.

Now that you have complete sets, consider the other subject types. Animals could include their pets. Inanimate objects could include the housing and physical items your refugees need. Connected organisations could include refugee charities, churches, mosques and temples, employment agencies, and the Home Office.

Using this mathematical process will help you to catalogue all the refugees' needs and assess the impact their presence will have upon the domestic community. With all this information at your disposal you can determine how many refugees you can realistically accommodate in each area, where best to house them, and what resources will be needed so you can create a comprehensive solution that works for everyone. Nobody will be left out of your planning because everyone is given equal consideration from the beginning.

Rules

Establish any rules connected with the experience or system you are analysing.

Applicable rule subsets may include regulations, laws, guidelines, traditions, religious practices, terms and conditions, codes of practice and government guidance, and more personal rules such as the principles of your subjects, cultural norms, their religious practices, and what they were taught for adulthood. These will be discussed in more depth later, (p49 p51). The set of language describing rules will include enforcement, evasion, avoidance, and loopholes. You may also find that there are no regulations for the experience, even though it is felt they are needed, or rules may be inadequate and require revision. You may create a rule or tradition to suit a situation.

Free will must also be considered at this point. People often evade or avoid rules. Therefore, you must consider not only which rules might apply, but the probability of them not being applied, and the consequences of people's free will.

Example: Ukrainian refugees

Consider now which rules will apply to your Ukrainian refugees. Are they religious, and if so, what religious obligations do they need to fulfil? Which cultural or social traditions do they ascribe to? What morals and principles are important to them?

Interview your refugees to discover the answers to these questions, and the extent to which they adhere to the religious and cultural rules of their society. Make no assumptions. Do not rely on generalisations. You need to dig deeper to reveal the truth.

Action

Consider the details of the experience and how it should be implemented. Conduct an audit to identify all your options and use your set thinking tools to order and prioritise them. Ask yourself, do you have a complete set of all the implementation methods available to you?

Example: Ukrainian refugees

Use your thinking tools to help identify the best way to organise the resettlement. You may need to form a team of people to assist you, for instance, arrange transport for your refugees, or arrange for a series of interviews with your refugee and host subjects. You may even want to run a publicity campaign to publicise your requirements for potential hosts and encourage people to volunteer.

Time

Consider when it will (or should) happen – the date, time, duration, timetable, and the timing of your experience. Are some dates more, or less, suitable than others? Will your project be delayed due to short staffing during lunch breaks, or affected by the school holidays? Ask yourself, do you need to connect with important dates such as religious festivals, bank holidays, children's holidays or birthdays, or Mothers and Father's Day? Be meticulous about the details and use set thinking to expand and order your ideas. Associated language here could include timetable, time frame, updating, short and long term, timing, up to date, latest, morning, afternoon, night, and lunchtime.

Example: Ukrainian refugees

Consider the best time to rehouse your refugees. Is the relocation time-sensitive, for instance? Or are there special or religious dates that need to be factored in? Think about how long the resettlement will take, how long it should last, and whether there are any milestones that you need to factor into your planning.

Place

Consider where your experience or system will take place. This may involve multiple locations – the places where your experience is planned, the places where it is implemented, and the places where it is evaluated. Ask yourself, what are the best locations to fulfil your needs, and have you considered every possible option? Or in other words, do you have a complete set of places?

Example: Ukrainian refugees

Consider everywhere that could accommodate your refugees. You could rehouse them as lodgers in family homes, in hotels, hostels, or in purpose-built accommodation, for instance. Classify, expand, order, and combine to find the best solutions. You will also want to consider in which regions of the UK your refugees should settle. Ask yourself, do some locations offer more resources, or less opposition to the plan?

Variable Consequences

As the name suggests, variable consequences relate to the fact that the impact your decision has will vary between individuals. These variable consequences can be gathered into sets, and can be explained mathematically as decisions = consequences, where both sides form a set. No two individuals feel or respond in the same way, so your decision may have multiple outcomes depending on the individuals involved. Consider this variety, it will help you to identify the truth. You can, if you choose to, also anticipate both the impact your subjects will have upon the world (such as the creation of additional waste), and the impact of the world on them (inspiring them to take an art class for instance). The consequences flow both ways.

Example: Ukrainian refugees

Imagine what effect your refugees may have upon the people, services, and businesses that are already there. Now consider what impact the domestic community might have upon them. Will they be accepted? Could they become the victims of racism or abuse? Will the refugees and their hosts get along, or will practical concerns or cultural and religious differences damage relations? Will community resources become strained, or could their residence provide benefits in terms of an influx of new skills, knowledge, or money?

The Secondary Classes of Knowledge

Objective

Setting the objective helps to concentrate the mind. The characteristics of the set which represents your system or experience should be defined (or restricted) by referring to your objective. Take time to consider what you are hoping to achieve. What does success look like? What would constitute a failure?

Example: Ukrainian refugees

Your objective appears obvious at first glance – to successfully rehouse the refugees in suitable accommodation. However, a closer examination of the goals of everyone involved reveals the situation to be far more complex. The government's agenda could be staying within budget, while the refugees may be most concerned with being housed near their fellow countrymen. You will need to take account of each of these objectives if you are to create a workable, and fair, solution.

Responsibility

Every action and object should have a responsible person allocated to it, with clearly defined roles. This may be several people. Delegating responsibility for set tasks, and taking personal responsibility for our own roles, are important parts of the process. For example, we could make an individual, or a group such as the local authority or householders, responsible for a tree – from obtaining the sapling to its planting, maintenance, and ultimately, its death. There may be various people responsible over the life of the tree, i.e., in a post allocated with that function. They could have a list or set of instructions.

Example: Ukrainian refugees

Let us consider who is responsible for making sure our resettlement goes without a hitch. What responsibilities should be attributed to the landlords, yourself, your support team, the government, the refugees themselves, and anyone else connected to the experience? Which members of your team are responsible for organising transport or allocating homes? What are the refugees and their hosts personally responsible for? Will they, for instance, be required to agree to a set of commitments, or adhere to certain rules of conduct?

Truth

Examine every piece of information rigorously, ensuring its validity and leaving nothing open to interpretation. There are many ways of doing this. You could use trusted sources such as experts in the field or those who have proved themselves trustworthy with this class of knowledge before. If this is your first time seeking such knowledge, you may need to rely on trial and error until you find the best source. You could also conduct your own thorough research. Sources could include bibles such as timetables, guidebooks, dictionaries, or research papers. In fact, any source that has been proven credible. Auditors, whose job it is to ascertain reality, use several of these methods in their work, regularly referring to original documents to check they have been told the truth and seeking independent advice and clarification.

Associated truth words could include tried and tested, experienced, or validated.

Example: Ukrainian refugees

Ensuring the truthfulness of the information you use to rehome your refugees is essential if your project is to be a success. Verify the information provided by your refugees and hosts to ensure there are no mistakes or intentional falsehoods. A failure to do so could result in refugees being placed in unsuitable accommodation.

Finance

Ask yourself, how much will the experience cost? Will it generate income or savings? Consider not just the numbers but whether it constitutes value for money, how it will be funded, and what budgetary controls are (or should be) in place. Including associated words can help you to compile a suitable mathematical model and could open your mind to other methods of funding. Funding the project could be achieved by grants, bequests, savings, loans, public donations, reserves, sponsorships, or be treated as running costs, for example. Other words associated with finance include profit, loss, bankruptcy, breaking even, value for money, costs, receipts, income, payments, expenditure, budgetary control, cost-benefits analysis, benefits, and investments.

Example: Ukrainian refugees

Your refugee programme may be financed by the State, or through public appeals or charities. Costs will likely include housing, school places, and operating costs. Also, investigate any financial benefits that may be achieved. Your refugees may bring new jobs into the area, for instance, or set up their own businesses. Their religious groups may encourage fundraising for local causes, or their presence may entitle the community to charitable or State redevelopment grants. These could offer significant economic advantages and contribute towards the project's success.

Information

Collating and sharing verified information is crucial to the success of any endeavour. The more informed you are, the more accurate your interpretation of reality will be, and the more information you share, the better-informed others will become. Such transparency will enable the spread of best practice and create a less ignorant world that is less susceptible to lies and biases. Information may also be necessary to keep the people in your project informed, for instance by sharing critical information in a crisis. Such information sharing can involve everything from signposts to leaflets, posters, TV adverts, letters, and the internet. In essence, any resource which informs people of facts that are relevant to them. Refer to your objective here. What purpose are you hoping to achieve and how can the sharing of information assist with this? Then share your process, results, and learnings widely and accurately so that others can trial similar projects and learn from your mistakes. Associated knowledge can include publicity, announcements, and giving information. The internet is a source of information.

Example: Ukrainian refugees

Ensuring your refugees are well informed about their rights, responsibilities, how to travel to the UK, and what to expect when they get here will help you achieve your objective – to integrate them seamlessly into the area. You should also share information about their arrival with the wider community, including the reasoning behind it, what measures have been taken to foster integration, and what benefits their presence will bring (as identified in finance).

Mistakes & Precautions

Consider what mistakes could occur, what precautions could be put in place to prevent them from occurring, and what actions could be taken in the event a mistake is made.

Mistakes could include everything from your sources being out of date or biased to customer complaints, addiction, and intentional errors such as criminal activity and scams. Risk assessments, management audits, quality assurance exercises, and safety reviews are useful tools here. Consider all the relevant mistakes associated with (connected to) the experience and what precautions can be taken to prevent a class of mistake or mitigate its impact. This could include double-checking your information with multiple sources or changing your procedures in response to upheld complaints. You will also need to plan your response in the event such mistakes occur despite your efforts. This could include emergency response planning or putting in place a system for evaluating mistakes and implementing additional precautions from the lessons learned. A quality assurance officer could prove extremely helpful with this, so consider assigning this responsibility to an appropriate person. Consider also what the consequences will be if and when mistakes happen. The worse the consequences are, the greater the prevention measures that will need to be taken.

Illnesses are a common type of mistake. Precautions can include any actions taken to prevent illness such as regular medical check-ups, treatments, inoculations, and a healthy lifestyle, as well as putting a system in place to learn from past mistakes. Set thinking can be applied here to cover all eventualities and identify which illnesses can be prevented and which cannot, such as inherited conditions. You can use set thinking to categorise all the inoculations available and to create a set of all the preventative lifestyle changes that could be made, such as eliminating behaviour X (smoking, for instance) and introducing behaviour Y (healthy eating, washing your hands and other hygiene activities, and looking both ways when crossing the road). Treatments can also be objectively evaluated, considering the ability of each treatment to either cure (X) or modify (Y) the condition, and those which have no effect (Z). You can also classify illnesses according to their seriousness, and their probability of occurring. Using set thinking in this way will enable you to put effective preventative measures in place - stocking your first aid box with the most relevant, and useful, items for instance. Now that you have a complete set of the illnesses that can occur, and an understanding of the most likely threats and most viable treatments, you can ensure that your first aid box is appropriately stocked.

Your precautions and response measures will need to be updated as necessary and your assessments widely shared, both nationally and internationally, so other governments, organisations, and individuals can learn from them and avoid similar mistakes. This is another process favoured by auditors, who frequently ensure controls are in place to prevent errors occurring in a system.

Let us consider a recent report by the UK's health ombudsman into sepsis management as an example. The report noted that too many patients are dying from this treatable condition because the NHS continues to make 'the same mistakes' it highlighted 10 years ago. Despite mistakes being identified, and a process being put in place to review mistakes as they occur (the ombudsman), effective precautions and responses were not implemented. Through the rigorous application of set thinking, this lapse could have been avoided. The NHS could have compiled a complete set of the mistakes that could be made with regards to sepsis, introduced precautions and controls to ensure best practice was carried out, and shared these in the form of a regularly updated manual of procedures which specified the mistakes that could occur, their corresponding precautions, and assigned responsibilities for each. By sharing mistakes and precautions in the form of mathematical models, we lay the foundation for a more transparent, and better organised, world.

Example: Ukrainian refugees

To prevent mistakes when rehoming your refugees, list and evaluate all the possible errors that could be made. For instance, an unsuitable host could be chosen for a particular individual or class of refugee, or their application may not be sufficiently comprehensive. Now consider which precautions could be taken to avoid this. You could compare applications to help you spot missing information, for example, have documents double-checked by another member of the team with a fresh pair of eyes, increase the number of questions you ask of your subjects prior to rehousing, or institute a trial period to weed out any personality clashes early on. You can also consider the actions that could be taken to remedy mistakes when they happen or lessen their impact on program success, for instance creating a list of back-up hosts and accommodations where refugees could be relocated to if necessary.

Probability

Consider the likelihood of each experience or event occurring and use this to tailor your plans. What is the likelihood of winning a competition for instance [aside I feel that any competition, especially TV, should show the odds], or of a natural disaster occurring? Language associated with probability includes stroke of luck, odds, and likelihood.

Let us return to our Primary Classes of knowledge for a moment. Consider how likely it is that each of the subject subsets you identified will occur. You will probably want to divert your resources towards those with the highest probability of appearing in your sample group, though it is important not to overlook the outliers and remain inclusive of minority subsets. Similarly, you will want to consider the probability of the rules connected with the experience or system occurring, and of each action, time, place, and consequence you identified.

This class of knowledge is particularly important to the identification and prevention of mistakes. You will need to systematically evaluate the likelihood of each mistake occurring to plan effectively. The more likely a mistake is to happen, the more care should be taken. This can also be applied to illnesses, by identifying which are the most common (such as cuts and colds) and those with a lower probability (such as kidney disease or rare infections). The demography of your subject groups (age and gender) will naturally impact your probability matrix as child subsets will be more susceptible to cuts and grazes while middle-aged men will be more likely to suffer heart failures and women will be more prone to gynaecological conditions.

Example: Ukrainian refugees

Let us return to the subsets we created for your Ukrainian refugees. Ask yourself, which of these subsets have the highest probability of being present in your refugee group so that you can tailor your project to their needs more effectively. Now return to the possible mistakes you identified and evaluate which are most likely, remembering to consider the demographics (age, gender, employment etc) of your refugees. You now have a clear idea of how and where to marshal your resources to prevent mistakes from happening and to respond to them effectively when they do.

Best Practice

Your experience should both adhere to best practice and contribute to it. Applying set thinking is a powerful strategy for seeking best practice. Before embarking on your experience, consider what the current thinking on best practice is and whether you can improve upon it. Use classify and expand to seek all available information on best practice through multiple sources and order them according to their usefulness for you. When your project has been completed, you can reflect on your learnings and by identifying and sharing best practice nationally and internationally help others to implement best practice going forwards. The sharing of best practice is important enough that it cannot be left to chance. Ensure the right person or organisation is allocated responsibility for the task. This could be an auditor, government body, or a central organisation of some kind.

Example: Ukrainian refugees

Consider, what is best practice for rehousing our Ukrainian refugees? Try contacting other countries who have run similar projects and break down the system they used to establish what mistakes were made, which mistakes typically occur together, and which precautions proved most effective. This can then be used to inform your project in the development stage. Allocate responsibility to a person or organisation for identifying best practice learnings from your own project, and ensure they share this information nationally and internationally so other, similar projects, can benefit from your achievements.

Quality

Evaluate the quality of the experience by comparing it to other similar experiences and how successful it was at achieving its goals. Was it comprehensive, efficient, and of a high standard? Were mistakes made, and how effective were the precautions and the response to these failures?

Example: Ukrainian refugees

Your refugees are now safely ensconced in their new homes. How did it go? Are your refugees content with the arrangement? Is the community welcoming them into their midst, or is their arrival presenting unforeseen challenges? Did you account for all the relevant sets, or were the needs of some of your subjects left unaccounted for? Were mistakes made, and if so, how well did your precautions and responses hold up under the pressure?

Your learnings could help other refugee resettlement projects go off without a hitch. If it went well, you might even become the gold standard for best practice.

Objective Thinking in Action

In this chapter, I wanted to share some of my own personal uses of mathematical thinking strategies. Particularly those which have inspired solutions to real-world problems which could positively impact our world. By doing so, I hope to demonstrate not only their ability to open our minds to unforeseen potential, but also to give you a greater understanding of how you can make use of them in your own daily lives.

The cost-of-living crisis

The challenges posed by rising prices and stagnating wages can seem insurmountable. They are certainly a source of anxiety and depression for many people nowadays. Mathematical thinking can, however, offer a solution to this emotional and psychological distress by providing a rigorous means of assessing all our financial options. By applying mathematical principles, we can identify all classes of finance and the members within each set, such as forming complete sets of costs (food, drink, mortgage and rent payments, utilities, petrol, TV, insurances, and toiletries), and income streams (pensions, salaries, and investment incomes). We can identify the classes of costs which are essential, or to express the point mathematically,

the costs which have essential characteristics, and prove it by considering the consequences of not incurring, or reducing, that expenditure. We will then be able to order our costs by value, concentrating our minds on the costs of most importance and helping us decide where to make savings. We can even create a set of how we can make such savings, including ideas such as using charity or second-hand shops, playing games or reading books that are free entertainment rather than spending money on entertainment, purchasing cheaper alternatives, recycling and reusing items rather than replacing them, and using trigger sets such as 'Refuse, Reduce, Recycle, Reuse' to help guide us in a more sustainable life. We can repeat the same process to evaluate our incomes, using classify and expand techniques to compile a set of potential income streams, such as selling items, working overtime, or getting another job, remembering to evaluate the consequences of our options to avoid negative results such as exhaustion or ill-health.

Footpath management

On a recent walk through the countryside surrounding my home, I passed over a small wooden bridge that had been erected over an area of bogland. It struck me that by applying a mathematical approach to the experience, I could gather all relevant knowledge about footpath management in my local area.

First, I created a set of the bridge's characteristics – the wooden strips and the nails that fixed them together – and used classify and expand to consider what alternative materials could have been used to build the bridge. I then created an upward set, describing the conditions under which (the characteristics) a bog bridge would be needed, and when it would not, considering such factors as how often the path was used and the above ground conditions.

By classifying the bridge as a wooden structure that people walk on, I was encouraged to expand the set to include all other such members, such as stiles, and bridges over streams and rivers until I had a complete set. I classified a stile as a way through a field boundary and expanded my set to include kissing gates, gates with stile steps over, and stiles which allowed access for dogs. There could also be classes of stiles which have the characteristics of requiring repair. I classified bridges over streams as a way of crossing a body of water and expanded my set to include metal bridges, stepping stones, and less official methods such as unofficially placed stones, and paddling.

There was a signpost on the bridge, which offered me an opportunity to create even more sets. I classified the signpost as an indication of the route for walkers and expanded it to include signs which contained the mileage to destinations, the footpath's name, and location, and directional arrows and finger pointers until I had a complete set. I realised that the path signs could also be treated as part of a set of signs with the same characteristics, such as height, width, and colour. I created an internal set of directional arrows, considering all its member elements such as its colour, size, material, and direction until I had another complete set. I also considered all the unwritten rules about arrow signposts such as their standard size, typical colour, shape, and material, and the direction it points towards (the path). I then did the same for signs with pointing fingers, considering the internal elements, including the option to have writing on the post (specifying location, direction, distance for instance) and unwritten rules for their usage. I then raised the classification to other signs. I considered all the possible classes of signposts in the countryside, those warning people of potential dangers, or instructing people not to litter, for example. This led me to construct a set of all the items carried by walkers which would later become classified as litter, and to connect with litter collection activities. This led me to create a new set, classified as ways litter could be collected, including organised litter collection activities by the council, volunteer groups, and by the individual walkers themselves.

I stepped off the bridge and onto a path, considering all the possible classes of footpaths – those in isolated situations on a road without any places to park, those with other paths nearby, paths crossing fields with crops, or through large grassy fields, paths through farms, and between houses. I analysed the footpath management system using the 15 classes of knowledge. *Reality* would include vegetation such as plants, trees, and flowers, including rare and invasive classes such as Japanese Knotweed. It would also include the human users, with varied subsets such as cyclists, horse riders, walkers, off-road drivers, the farmers, and landowners, rescue personnel and vehicles, and those users with physical disabilities – all of whom could be catered to (connected to) or not by the designers of the footpath. There could also be animal path users, such as dogs, wildlife, and livestock, and inanimate members such as the walkers' shoes, and rainwater. I also connected with relevant organisations such as walking groups and conservation volunteers. This led me to classify groups which could help with the management of the path, and I expanded the set to include nature site volunteers, and wildlife charities.

I reviewed the *actions* involved (walking, repairs, inspections, and maintenance), the *time* (when the path was utilised – during the day, at night, at certain times of year), and the *place* (where the signs, paths and bridges were situated). Now I considered the *rules* in play, establishing which rules applied – country codes, byelaws, council tradition – the extent to which free will might influence their application and the potential *consequences* of not abiding by these rules. I analysed additional consequences too re reality, the effect of the walkers, plant growth, the weather, and animals on the path, signs, and bridges.

Moving on to the secondary classes of knowledge, I considered the possible *objectives* (accessibility, encouraging usage, and preventing littering and other damage to the natural environment); the information that required *sharing* (instructions such as to keep dogs on leads, close the gate, and not to feed the animals, and knowledge sharing such as directions, maps of the area, and books detailing local walking routes that included footpaths); *mistakes* (obstructions such as fallen trees and foliage covering the signs, damage to the paths, signs, and bridges, the footpaths falling into disuse, fires, and users suffering an injury), the *precautions* that could be taken (regular inspections, encouraging walkers to report faults, either closing unused footpaths or better advertising their availability, first-aid resources, and prohibiting barbecues) and their probability (identifying the most likely mistakes and focusing on their prevention); *best practice* (ensuring clear pathways and signage); *truth* (ensuring the accuracy of signage); *quality* (well-built and maintained footpaths and clear signs); *finance* (the cost of building, repair, and maintenance); and *responsibility* (assigning a designated footpath officer to oversee the footpath network and proscribing personal responsibility to the walkers themselves).

By applying set thinking to my walk, I had, without even intending to, gathered all the relevant knowledge I would need to create a comprehensive, effective plan for managing footpaths – not only in my area but across the country. If local councils were to apply such mathematical rigour to their footpath management strategies, they could create a cost-effective system that works for everyone who uses it.

Diversifying farming practices

A few days ago, I was watching an episode of the BBC show Countryfile that described how farmers were using sheep fleeces to create mulch to surround vines, a practice which had transformed an unprofitable product into one with reasonable returns. As always, my mind instinctively applied my set thinking tools and 15 classes of knowledge to the new information. I began creating a set of languages describing fleeces with the certain characteristics that made them suitable and profitable for mulching. I applied it to the set of vines in the vineyard, considering whether its usage needed to be restricted to vines with specific (X) characteristics only. After describing the characteristics of the fleece, I asked myself whether there were any other profitable uses for them (other members of the set), such as using fleece to fill sofas. I then raised the classification to include all waste products on the farm, considering whether profitable alternative uses could be found for them too, such as using fallow fields for solar panels, or creating sustainable habitats for wildlife. I then raised it again to include all vineyards in the UK with Y characteristics, for instance, farms with sheep for whom it would be profitable to sell fleece as mulch, and then internationally.

Other possible sets then came to mind. As this is an example of best practice, it occurred to me I could create a set of all the best practices identified in this episode of

Countryfile. I could even raise the classification to include all best practice ideas demonstrated in Countryfile episodes, or all BBC programmes, or all television programmes in the UK. I could refine my analysis, limiting my set to best practice ideas in farming, for instance, or technology, or anything that suits my purposes, and assign responsibility for spreading best practice through the sector to a suitable individual. Through this method, I would then have created a complete set of best practice ideas on a topic that could be shared internationally to create a blueprint for societal endeavours. Alternatively, I could classify and expand a set of items which are normally unprofitable, but which have the potential to generate income if put to other uses – just like the fleeces had been. Unused fields, for example, could be transformed into solar farms. I could then raise the classification to include any organisation or person seeking such solutions for unprofitable items.

Set thinking, applied to something as routine as an episode of my favourite TV show, had conceivably led to the creation of a workable plan with real-world societal benefits. If only everyone was thinking in this way, imagine the accomplishments we could make.

Teaching Children Set Thinking

Summary

Our society places far too much emphasis on gathering knowledge, while paradoxically, leaving the acquisition of the most crucial knowledge of all – that which a child needs to know for adulthood – to chance. A new curriculum must be created, with international agreement, to instil these skills and understandings within future generations. In this chapter, I outline the most crucial lessons for adulthood identified by my research and demonstrate how these can be taught to children.

Today's children are tomorrow's politicians, scientists, artists, and business leaders. They will shape the future of our society just as completely as we shape them through our parenting and teaching. Our world is facing existential threats, and we must prepare our children to tackle them head-on by instilling in them the learnings they most need to help us create a more equal, inclusive, and peaceful society.

We must rid ourselves of the notion that more knowledge is invariably better. This is a fallacy. Our societal obsession with furthering our children's acquisition of knowledge, most of which is unnecessary or incorrect, is intrinsically damaging to our youth. As if this were not bad enough, we then, paradoxically, leave the acquisition of the most crucial knowledge of all – what a child most needs to know for adulthood – to chance. We rely on parents to instil these skills and understandings in the next generation, offering them no guidance or framework to steer their efforts beyond what they themselves were taught as children. Schools have become a place to digest facts, figures, and dates that clutter our children's minds and overwhelm their battered senses, when they should be reinforcing the more useful information passed on to them by their parents, such as the skills needed to socialise and connect with others, to anticipate likely mistakes, and skills that are relevant to everyday life. We teach our children to pass examinations when we ought to be teaching them how to live well and make a positive contribution to the world.

It is time to ask ourselves, how much of the information we download into our children's minds is really of value? We must differentiate between knowledge which will help them live better, happier lives, and that which is nothing more than an unwarranted distraction from achieving their full potential. This is not to say that learning cannot be a joyful pursuit, or that we should attempt to stifle our children's drive toward discovery. We should not undervalue the importance of learning for its own sake, whether it be history, physics, or the simple pleasure of reading a good fiction novel. It is also undeniable that certain knowledge is necessary for achieving life's goals, such as building a career or attending university. But our education system is weighted too heavily towards the production of tomorrow's workforce, and too little on teaching children what is useful for living their lives.

This is not a project we can entrust to individual parents or teachers. We require a society-wide agreement on which information is beneficial, and which is not; a blueprint of lessons for adulthood that is discussed, agreed upon, and formally recognised at the highest levels of our societies, both nationally and internationally, so that an education system can be created that emboldens our children, rather than hobbling them. Parents too will need to be better informed about the lessons for adulthood they ought to be teaching their children, so the blueprint must be disseminated widely, in the form of a how-to guide or website for parents, so that every child, no matter their background, can benefit from them.

In this chapter, I will endeavour to familiarise you with those lessons for adulthood which my research has highlighted as being of most importance. These should be standard in my mathematical world and can be expressed as a set with the characteristics of something a person needs to know for adulthood. This is not a comprehensive list, far from it, but it is a place for our society to start building a curriculum for learning which will empower our children and teenagers, equipping them with the skills and mindsets to one day lead our society into the positive future we aspire to create.

Lessons for Adulthood

Objective Thinking Strategies

The most important lesson for adulthood we must teach our children is mathematical thinking. Objective thinking strategies lie at the heart of what children need to know to lead full and successful adult lives. The core tenets of this form the basis of this book: not seeking reality from information received directly through the senses (Lesson 1, p13), and to concentrate on living their lives with set thinking and the four thinking tools (Lesson 2, p18), and the 15 classes of knowledge (Lesson 3, p30). These act as a foundation, upon which all other lessons for adulthood can stand, and will be truly life-changing for our children. Emboldened by their new skills for objective thinking, children can make the most of their lives, appreciating every moment and making discoveries about themselves which will enrich their existence.

Teaching our children set thinking encourages them to think through the day-to-day decisions of their lives, properly evaluating their choices so they make good decisions, and contemplating the consequences. It also encourages them to make life choices based on reasoned judgements. By recognising the reality that they experience is a set of knowledge to be investigated, they learn to plan and prioritise.

They begin to think ahead, laying out their clothes for the next day, for instance, or making sure they have enough money for their bus ride to school, and to order their actions so the more important tasks are completed first. They get in the habit of making connections, expanding their thinking to consider new and unexplored possibilities. And they make the best use of their time, filling in free time with productive activities.

Every aspect of their experience can be classified and analysed using set thinking, from the items on their dinner table to their journey to school. This lesson should be instilled at the earliest possible point in their development so set thinking becomes ingrained, and they grow up viewing the world through mathematical principles. But failing that, it is never too late to begin teaching our youth to think using mathematical principles. Indeed, the benefits to teenagers of applying mathematical thinking are so great that it is imperative we integrate set thinking into the high school curriculum as a matter of urgency to protect our youth from delinquency, juvenile crime, antisocial behaviour, and mental health issues. Remember, children are mathematical creatures. Set thinking is in their nature, so it will be far easier to suppress their slide towards subjectivity than it will be to overcome our own. To help you with this, teaching aids are provided later in this book (p59).

Living in the Moment

We must teach our children to take each experience one at a time. To live in a bubble, without seeking to understand the elusive, incomprehensible, reality of their experiences. To concentrate on living their life mathematically away from the troubles of the outside world. To start afresh in each moment, without reference to the past.

Learning to move on is crucial to their ability to live mindfully, unburdened by the history of their own lives, or by societal memory. We must, therefore, teach our children to focus on the here and now, living each moment to its fullest, and take from the past only what is necessary to learn from past mistakes and make well-informed choices.

Teaching our children not to create perceptions will offer protection to their fragile minds, acting as a deterrent to the mental health problems, suicidal thoughts, and the antisocial behaviour that is so often a cry for help from today's overwhelmed youth.

Socialisation

Teaching our children how to socialise is paramount if they are to lead happy, productive lives. It offers a means of connecting (or forming a set) with other people, and is the key ingredient for forming friendships, attachments, and lasting commitments. It also prepares them to function within the societal parameters of their communities. Yet our education system barely acknowledges the skill at all. It certainly offers no formal training on the subject, despite freely punishing children for the poor behaviour that typically results in its absence.

It is left to parents to teach their children how to interact with others, how to be polite, to share well, and become good conversationalists. However, not all parents are equally equipped to teach children such skills. If their own socialisation training in childhood was lacking, then these poor skills will inevitably be passed on to their children, perpetuating the cycle. We therefore need a formal system for socialisation training that includes such techniques as how to use small talk (asking about the weather or discussing their interests for example); agreeing or connecting with others on subjects even when you disagree or do not understand the topic; adopting manners by saying please, thank you, sorry, excuse me, hello, and goodbye; how to connect with the theme of the conversation, or if desired, change its direction; and how to emotionally engage with others, for instance by asking after their wellbeing. We should also be teaching children how to connect with other people (make friends), for instance by sampling new activities to determine what they enjoy and meet like-minded people and learning to think about others feelings and viewpoints by putting themselves in their position.

This training ought to instil in our children an understanding of the unwritten rules of our society – such as saying hello to strangers who acknowledge you in the street, or thanking the bus driver as you disembark – and their importance as effective tools for connecting us with other people. Politeness is not an inherent skillset. It must be learnt. Unfortunately, for our children and our society, our current system for passing on this knowledge is wildly insufficient. Thankfully, social interaction contains mathematical elements. Set thinking can therefore be applied. When you connect with another person you are creating a set with that individual. These mathematical elements can be learnt and then repeated in a variety of social contexts, manners for instance. Each time a situation with X characteristics arises, you say thank you. When a situation with Y characteristics occurs, you say sorry.

Tolerance, Empathy and Understanding

We must teach our children that while everyone is different, we are all based on the same template. This is self-evident when you apply set thinking to the way we view other people and ourselves. By classifying and expanding on a person, or group of people, we discover that while everyone is unique, they are at the same time, inherently the same. They form a set. Through the teaching of an objective thinking strategy, therefore, we can instil in our children an awareness of the ways in which we are, all of us, alike. This encourages tolerance, empathy and understanding, helping them to put themselves in other people's shoes, irrespective of race, creed, or social status. Without generalisations and stereotypes to cloud their judgements, they can see people as they really are. This is crucial learning for the next generation. Rather than burdening them with our own prejudices (both conscious and unconscious) we can prevent them from perpetuating our mistakes. They will be free to create a future society that is not polluted by racism, sexism, and other forms of bigotry (p78).

Knowing themselves

Children must be taught to seek and identify patterns in their own behaviour. Through this process of self-reflection, they will learn to better understand their own identities. They can also examine their parents' behaviours as a guide to understanding their own personality and character, and reflect on the behaviours of their friends, teachers, and other adults in their lives. An objective thinking strategy will enable them to examine each variable in turn, considering in what respects they are similar and dissimilar to other people.

When our children know themselves, they become better equipped to make up their own minds. Rather than blindly following the path laid out for them or copying the behaviours and choices of other people, your children will ask, what do I like? Is this person worth following? Do they have expertise on the subject, or a proven record of reliability? Will this activity be good for me or make me happy? This is the key to independent thought, and a life lived on their own terms.

Discovering their interests and talents is a key part of this process, particularly those non-academic abilities which are so often left unexplored by our education system. This could include practically anything that appeals to them - sport, art, photography, singing, bird watching, or chess. Such activities could also provide a means of connecting with people, or in other words, making friends. A key learning goal for our children should therefore be to identify these interests and explore them to their fullest extent. Only then can they reach their potential.

Set thinking offers a useful tool for this. It can encourage children to explore new possibilities and try new things, widening their horizons and helping them to discover more about their character. Let us imagine, for instance, that your child plays football at school. By creating a set of sports which they could participate in they can expand their options out to include rugby, tennis, badminton, and karate, perhaps sparking new interests and encouraging them to sample new activities to identify which they connect with. Similarly, by creating a set of musical instruments they could learn to play they are encouraged to imagine beyond the strict confines of the trumpet lessons their friends are taking. Who knows, you may have a classical cellist on your hands.

Evaluating Themselves, and Others

We must teach our children to use set thinking to properly evaluate themselves, and others. In this way, our children can learn to value those characteristics of a person which matter most, rather than using subjective measures of worth which all too often result in low self-esteem and faulty judgements.

Subjective critiques are responsible for many of the preconceptions and inanities that underpin racism, prejudice, celebrity culture, and bullying in our society. Our children learn their concepts of beauty, desirability, and popularity through mirroring the evaluations we make, or that they see on TV and social media. This shapes how they feel about themselves and about other people. They denigrate themselves for not living up to the standards of perfection they are encouraged to set, and in turn, both resent other people who do, and judge others who do not. It need not be this way. When you review the set of knowledge that represents a person and connect or rate each quality (ordering) you inevitably find that the most important qualities are those not directly discerned by the senses. A person's appearance, body image, and smell are attributed far less significance when analysed and critiqued using mathematical principles than those aspects which require greater attention to notice, such as their helpfulness, kindness, loyalty, and honesty.

In discovering this, our youth learn something important about themselves. That their own physical appearance, body image, and how they are perceived by others is insignificant compared to who they are, and how they contribute to the world. This does not mean that they should not take care over their appearance, which could denote a lack of respect for themselves - or in certain situations, for others e.g., parents. Rather, they will learn to appreciate the triviality of societal notions of beauty and vanity (p78). This life lesson will shape them into far healthier people, in mind, spirit, and body.

Common Mistakes

We must teach our children which mistakes they are most at risk of making, and the precautions they can take to avoid them. These should be taught together, to instil in our children the preparedness they need to stay in control of their lives. The most common mistakes should also be formalised, included in the agreed lessons for adulthood guidance given to parents and teachers, and continually updated to remain inclusive and relevant.

Make use of the 15 classes of knowledge to identify which types of mistakes your children have the highest probability of making and focus your teaching on those most of all. These could include losing things, not looking both ways when crossing the road, unhealthy eating, and falling prey to scams. We can lessen the impact of these mistakes by teaching precautions such as to check they have everything before leaving, to ask a teacher or parent before handing over their money or possessions to someone else, how to read the nutrition labels on food, and the Green Cross Code.

I also want to stress one, crucial, aspect of this lesson. The example we must set, as their parents, teachers, and guardians. The better world we seek is dependent on our children's ability to be open and transparent about their mistakes; their willingness to admit to failures and learn from them. This starts with us being open about our own. When they make a mistake, respond with empathy rather than criticism. Encourage them to consider where they went wrong, and how to avoid making the same errors in the future. They must learn that mistakes are not something they will be judged by. Instead, they are an opportunity to learn, and to share these learnings with others.

Routines

Routines provide children with a much-needed structure and help ensure vital matters are carried out. They are also important in the way society is organised and so are an important area of knowledge to share with our young. Each routine can be expressed as a set of knowledge.

While having routines is important for children, so is teaching them the need to evaluate their routines and consider alternatives using set thinking. Not doing so could lead to them becoming stuck in a rut and failing to seize the opportunities available to them. Equally important is free will. We must teach children that their routines need not be rigid or unchangeable. At any point, they can choose to alter their routines, assuming, of course, they are willing to accept whatever consequences arise as a result. Set thinking can help them to identify these too.

Each routine forms a set which can be evaluated using set thinking tools to guarantee they continue to be fit for purpose. Consider, for instance, a child's morning routine. They may get up, get dressed, wash, brush their teeth, and eat their breakfast. Applying set thinking to this routine encourages children to consider the consequences of performing it, or not, and presents alternatives. What would happen if they did not brush their teeth thoroughly? How much effort is required and time taken up? Ought they follow their dentist's advice to brush after eating their breakfast, or do it the other way around? Do they need to shower every morning? Must they wear their school uniform? Thinking through the alternatives, and the consequences, helps children learn to make the correct choices.

Consequences

Teaching children to examine the consequences of their decisions, from the most insignificant decision of what to put on their porridge in the morning, to the most life-altering choices they experience, is essential. By teaching them to factor consequences, where possible, into their thought processes on small decisions, we give them the tools and inclinations to do so with the more serious matters and encourage them to take personal responsibility. Objective reasoning in one area of their lives inevitably bleeds into them all.

This ties directly into our parenting and teaching strategies. Often children are taught not to behave in a certain way because they fear 'getting in trouble' or being penalised – missing out on a treat or losing their pocket money, for instance. This form of behaviour modification may prove successful at preventing certain undesirable actions, but it will not teach children to master reasoning and rationality which will help them make better decisions in the future. For that, they need to understand the reason that behaviour is forbidden, and the consequences that could arise if such behaviours go unchecked. Instead, give children a justification for why that behaviour is not permitted and explain the impact it could have.

Let us take swearing as an example. We routinely admonish children for using 'bad' words, and they quickly learn which words are acceptable to use and which are not. But have you ever explained to a child why they should not swear? It is notable that while our children successfully learn that swearing is bad, most grow up to use swear words. We have succeeded in teaching them not to swear as children, but the lesson only lasts until there is nobody left to punish them. If instead, we encourage children to think through the impact swearing may have on those around them, with our help of course, the lesson is more likely to stick. They realise that swearing may upset other people. That some may find it offensive, or even harmful. That it may make other people think badly of them. That swearing is typically unnecessary and more socially acceptable options are abundant, and that by using other words instead they may appear more intelligent, more thoughtful, or more considerate. A far more impactful life lesson than simply losing their TV privileges for the evening.

Rules for life

Every culture has its own unique set of rules and norms that its members are expected to live by. Children should be familiarised with the most relevant of these, so that they can adapt and fit into their society.

Rules take many forms (p34). Some may be arbitrary or custom-specific such as table manners, religious habits, or personal principles, while others instruct people how to live their lives healthily and safely, such as hygiene standards, chewing before swallowing, sex education, or the Green Cross Code. A few may be legally proscribed, such as tax laws and financial regulations. In each case, the child must be taught to determine which rules apply to each experience. Be mindful, however, of not overwhelming children with more knowledge than they need. Focus on teaching those rules which are necessary for their place in society, and at this stage of their lives. They will also need to learn an appreciation of free will; an understanding that although these rules apply in their society, it is up to them to decide which rules to abide by, and which to ignore. Not following the rules can, however, have consequences, both large and small, so children must be educated on what these may be so they can use set thinking to make an informed choice.

Teaching Aids

Teaching set thinking begins with instilling in children an understanding of what constitutes a set. This can be shared with children as young as 18 months old. If they can understand language, they are old enough to understand the basics of set thinking.

A set of coloured brick toys is a perfect learning tool for this, and one parents often already use to teach children colours. As you point out the yellow, green, and red bricks, you can use them to instil the understanding that these items form a set. You can make this clear by introducing other toys into the game to demonstrate they are not part of the same set - a cuddly toy perhaps, or a musical instrument. A simple game of 'spot the odd one out' is a great learning tool for teaching set thinking. Now add several other toys into the pile. You now have another set – a set of the child's toys. They become connected again, and by adding something else to the pile, your keys for instance, you have your next game of 'spot the odd one out.'

As children age, you can begin to teach them more complex mathematical concepts. The same set of coloured bricks can be used to teach children the concept of a complete set. This is the state of knowledge we must aspire to achieve with our reasoning. Lay out several bricks of the same colour – let us say you now have a pile of red bricks. These bricks form a set, classified as red bricks that the child owns. But is it a complete set? Is every red brick the child owns in the pile?

Taking the analysis further, we can conclude that other children may also have red bricks and that therefore their bricks are part of a set of all the red bricks owned by their friendship group, or in the whole world (an upward set). We can also encourage the child to catalogue all aspects of the bricks, for instance that it has 6 sides (a downward set). We can keep doing this, raising the classification, and classifying and expanding, until we have a complete set. We can also repeat this process with the other coloured bricks the child owns, classifying a set of all the green, blue, or purple bricks the child owns. We can also teach this lesson using our set of toys from earlier. Consider, are all the toys your child owns included in the set (a complete set)? If not, encourage them to gather all their toys until the set is complete.

You can teach set thinking using any object. As all objects can form a set, you can tailor your teaching to whichever one piques the child's interest, farmyard animals for instance.

Let us use Moo Moo the cow.

First, we can create a downward set, noting that Moo Moo has a pair of legs, a head, a body, a tail, pointy ears, and black spots. We can create another set from Moo Moo's head, noting that it has two eyes, two ears, and a snout.

Next, we can create an upward set. Moo Moo is standing in a field with many other cows, so she is part of a set of all the cows in the field. We can now raise the classification. Farmer Jack has lots of other animals on his farm – pigs, sheep, and horses, so Moo Moo is part of a set of all the animals on the farm. There are lots of farms all over the countryside and Moo Moo is part of a set of farm animals all over the country, and all over the world.

Of most importance is to stress to the child that Moo Moo is connected to them all, and all of them are connected to her because they are alike (or in grown-up language, they share characteristics).

You can also teach set thinking using the child themselves. Create a downward set, noting that they have hands, feet, eyes, hair – the list goes on. Then create an upward set. This could be their family, or perhaps the children in their class. Then you can raise the classification to create a set of all children in the school, then all children at schools in the country, and ultimately, all children, everywhere.

Averting Disaster with Set Thinking

Summary

Many of the worst disasters in history were caused, or made incalculably worse, by poor planning, avoidable mistakes, and bad decisions. In our subjectively organised world, disasters are inevitable. Mathematical thinking holds the key to protecting ourselves from future calamities. In this chapter, I demonstrate how and show how three modern-day disasters could have been avoided with objective thinking strategies – the RAAC crisis, the Piper Alpha disaster, and Hurricane Katrina.

What is history, but a catalogue of events that we could learn from? Tales of failed endeavours, disastrous decisions, and calamitous outcomes; Glorious victories, heroic saves, and wise leaders. Such stories contain within them all the information we would likely ever need to learn from our forebears' mistakes and stop them from happening again. Every one of their successes is a roadmap towards our own. Yet, we relegate them to the history books, choosing to ignore the wealth of knowledge they offer. And so, history's failures are repeated, time and again.

Mathematical thinking offers us an escape from this endless loop. By applying set thinking and the 15 classes of knowledge, we can learn from the mistakes of the past, develop effective precautions, and share best practice, protecting ourselves against future disasters. Armed with all the relevant connected knowledge, we can create more effective, safer systems, and with a range of mathematical tools at our disposal, we can respond swiftly and with purpose when disaster strikes.

Unfortunately, my calls for a mathematically organised world have not yet been heard. With subjective thinking the driving force of our society, tragedies are inevitable – a predictable consequence of poor planning, erratic responses, and ill-informed decisions. Many of our worst disasters have arisen as a direct result of failures in risk management. Organisations ignore the need to properly assess the mistakes that could expose them to catastrophe. They avoid looking too closely at the failings inherent in their designs and processes, preferring to believe themselves invulnerable. When the worst inevitably happens, they are left unprepared for the fallout, rushing to find last-minute solutions.

This unwillingness to predict and plan for mistakes is sometimes the result of cultural bias – arrogance, overconfidence, or ego. Other times, it is a consequence of greed, with companies accepting high risks in return for large profits - a strategy shockingly exposed by the Ford Pinto scandal of the 1970s in which a structurally unsafe vehicle was brought to market when a cost-benefit analysis showed it would be cheaper to pay settlements for deaths, injuries, and burnouts than to repair the flaw. Often, they are spawned by fear. Imagining worst-case scenarios is, after all, an uncomfortable endeavour. But whatever the cause, the consequences can be catastrophic – businesses fail, lives are lost, and economies crash. Paul Moore, former head of group regulatory risk at banking group HBOS, famously dubbed the 2008 financial crash 'the biggest quality failure of all time,' and a direct result of excessive risk-taking by the banking sector.

Mistakes and precautions are just one of the 15 classes of knowledge that can play a crucial role in averting disasters, and responding to them when they occur. By assigning the right responsibilities to the right people, and ensuring every process has a person to oversee it, we can help ensure important information is not overlooked and be in a stronger position to act in a crisis (responsibility). By sharing information and best practice, we can contribute towards the implementation of better, safer systems.

In this chapter, I will explore three modern-day crises which could have been averted with the use of mathematical models – the RAAC scandal, the Piper Alpha disaster, and Hurricane Katrina. Perhaps by applying set thinking and the

15 classes of knowledge to them now, we can at least learn enough to prevent their like in the future.

The RAAC crisis

In 2023, the UK was rocked by a scandal. News broke that more than 150 school buildings were on the verge of imminent collapse, endangering the lives of thousands of British children. Schools were closed, teachers scrambled to erect temporary buildings to educate their students, and accusations of government mismanagement, cover-ups, and indecision ran rampant. The cause of the structural failures was a building material - reinforced autoclaved aerated concrete – regularly used in civic construction projects between the 1950s and mid-90s. Lightweight, strong, and cost-effective, the material only had a 30-year lifespan – and its time was up.

Parents, teachers, and the media demanded answers. Why was such a short-lived material chosen in the first place? Why were there no plans in place to replace it when the time came? And what was going to be done about it now? Perhaps most worrying for those affected was the seemingly chaotic and incoherent response of the government to the crisis. They waited months (some reports say years) before announcing the problem, leaving teachers no time to prepare, and dithered over who would have to pay for the repairs (the government, or the schools themselves). When the news finally leaked, they scrambled to compile a complete list of at-risk schools, and seemed to collapse into disarray when it came to light that a succession of other public buildings, from hospitals to law courts, would have to close due to RAAC construction of their own.

What struck me most when I first read about this crisis was how easily it could have been avoided with proper planning. Using RAAC in school construction was an obvious mistake, its consequences foreseeable, and its outcome predictable. If the architect of this policy had used my mathematical models, they would have seen this disaster coming. Foresight trumps hindsight every time.

By using set thinking to compile a complete set of mistakes which could be made in this class of situation – conducting risk assessments, management audits, quality assurance reviews, and the like – they could have predicted not only that the RAAC would fail, but when it would fail, and what the consequences of that failure would be. They could have taken precautions, such as using a different building material, or putting in place measures to replace the material in an orderly fashion when it expired. And they could have shared their findings (information), nationally and internationally, so that those involved in similar construction projects would be forewarned.

A review of the primary and secondary classes of knowledge also reveals multiple other opportunities for the problems with RAAC to have come to light during the planning phase – while compiling a complete set of all the consequences of the policy; while considering its objective (which presumably would have included the need for the building to be safe, fit for purpose, and long-lasting); while assessing its finance (including the cost of repairs and maintenance); while assigning responsibilities (including building safety); while considering best practice (did this building material cause problems previously?); and of course, while assessing its quality. Unfortunately, without the benefit of an objective thinking strategy, none of these opportunities were taken.

The government could also have benefited from set thinking when planning their response to the crisis. A mathematical approach would have ensured the gathering of all connected knowledge in a systematic way. Rather than a panicked, slapdash hunt for information on how many schools were affected, classify and expand tools could have been used to create a complete set of all schools in the UK with RAAC construction. By then raising the classification to other types of buildings in the UK with RAAC and expanding to include hospitals and other public buildings, they would have discovered that the crisis was not limited to schools in time to take assertive action. And by creating a set of all the available solutions to the crisis and using set thinking tools to order, prioritise, and assess them, they could have selected a workable solution in a timely manner.

Over the coming years, this scandal will no doubt continue to be investigated. A public inquiry seems likely. The question remains, however, will the lessons identified by this report be used wisely to inform the planning of future projects? Or will they be read once and then forgotten, like so many public inquiries have been before?

Piper Alpha Disaster

One of the deadliest industrial accidents in history, the explosion of the Piper Alpha oil platform in July 1988 killed 167 people and caused devastating environmental damage.

Subsequent investigations identified the cause to be a startling array of human errors and organisational failures – mistakes that could have been easily prevented with mathematical thinking. The disturbing truth is, if set thinking tools had been available to Occidental Petroleum executives at the time, this disaster would never have taken place.

Reports into the disaster have shone light on a litany of mistakes: Maintenance crew failed to properly communicate information to each other, resulting in the use of faulty equipment; staff were poorly trained in how to respond in an emergency; production was prioritised over and above safety; there were a lack of clear procedures for shutting down the pumps in the event of a fire; the oil platform was poorly designed, leaving crew vulnerable in a crisis; and safety reviews were infrequent, and their findings largely ignored. Taken together, they amount to a partial set of all the mistakes that could put the oil platform at risk. If only an objective review of all the connected knowledge had been conducted prior to the explosion, it could have identified the mistakes inherent in the system, rated the likelihood of their occurring (probability), and established a complete set of preventative measures. Instead, the organisation fell afoul of its own complacency.

It is a tragic indictment of the way our society functions that we put so much effort into compiling sets of mistakes after a disaster has occurred, and so little effort into doing so beforehand, when they could achieve the most good.

Hurricane Katrina

In 2005, Hurricane Katrina made landfall on the Gulf Coast of America. New Orleans was buried underwater, 1,800 people lost their lives, and $100 billion of homes, businesses, and infrastructure were destroyed. Myriad system failures soon came to light, which proper planning and preparation should have caught, and dealt with, prior to the disaster. Two-thirds of the flooding was caused by multiple failures of the city's floodwalls, which proved to be deeply flawed. A management audit which included a proper risk assessment and quality assurance would have captured these mistakes.

But a failure to identify likely mistakes, and take adequate preventive measures, was just the tip of the iceberg. The recovery effort was hamstrung by a series of systemic failures. The 2006 bipartisan house report on the disaster, 'A Failure of Initiative', found that federal agencies were unfamiliar with their roles and responsibilities under the National Response Plan and the National Incident Management System, hampering their ability to evacuate the city and help survivors; an issue that would not have arisen if the 15 classes of knowledge were utilised, and responsibilities were properly assigned and communicated to all parties. A breakdown in communication between the various agencies on site also paralysed command and control during the emergency, and handicapped attempts at cooperation. Lives could have been saved if only a mathematical approach to organising the recovery effort had been taken.

Reorganising Society

Summary

The global adoption of mathematical thinking strategies will have far-reaching implications for the way society is organised. It will influence our institutions, our organisations, our politics, and our social interactions. As our society reinvents itself to accommodate our less subjective character, we can begin to effectively tackle the numerous threats facing our world, such as war, climate change, and pandemics, in a far more effective, rational, and inclusive way than ever before.

Mathematical thinking has the potential to transform more than just individuals; it can reorder society itself. Our institutions, organisations, and governments can all be reconceptualised, restructured, and better managed using mathematical models.

Imagine a world where best practice is shared across nations. Where mistakes are acknowledged, shared, and learnt from. Where decisions are made based on facts, rather than ideological assumptions. Where consequences are given more weight than expediency, and angry debates and rhetoric are replaced by co-created mathematical models. Objective thinking can make this possible.

A fact-based approach, that captures all relevant knowledge and examines it through a mathematical lens, will create more resilient, effective solutions to the world's problems. By considering every facet of the situation, anticipating likely mistakes alongside the precautions we can take against them, sharing best practice, and setting clear goals without interference from subjective generalisations or preconceptions, all systems can be improved upon and organised more effectively. Set thinking provides us with the tools to identify every problem to be addressed, consider every eventuality, and be aware of every consequence. It also enables us to identify and connect with all the different classes of people that could be impacted – whether they be rich, poor, disabled, or star athletes. This newfound inclusivity will ensure that nobody is left behind and the best option is chosen, for everyone, each time.

Let us consider some common system faults which are intrinsic in our modern world against their mathematical solutions. Firstly, a system may be piecemeal. The 15 classes of knowledge and set thinking can overcome this by gathering all connected knowledge for the system or experience.

Secondly, it could lack consistency. Set thinking can address this too, as by creating an appropriate set (where the characteristics of the set is everyone correctly doing X in a certain way) we ensure that everyone is singing from the same song sheet.

Thirdly, it may lack coordination. We can eliminate this by using our classify and expand techniques to ensure we have a complete set and that rigorous processes are in place. For instance, when identifying the relevant bodies to connect or contact in the event of a particular event (let us say a crime being committed), we can classify a body to connect with and expand to other bodies until we have a complete set.

Fourthly, it may not have been fully thought through. Connected thinking can help prevent this by compiling complete sets for each variable, thereby ensuring all aspects are fully considered, and identifying consequences, mistakes, rules, and best practice. It also encourages the sharing of information, which is often lacking in capitalist run societies which favour competition over collaboration to protect their intellectual property and boost profits. It should be noted the system may exclude sensitive information for sharing.

Finally, the system may not be comprehensive. Again, by checking that we have complete sets, with all members identified, we overcome this problem. Mathematical thinking can therefore be used to overhaul every system in our society so that it is functioning optimally and effectively for everyone who uses it.

Objective reasoning can be applied to develop a solution to any organisational issue, from determining how much to pay employees to allocating government funding. Consider, for instance, how one could go about deciding whether to offer a pay increase to NHS doctors. Rather than becoming enmeshed in ideological arguments on the issue, objective thinking strategies can be applied to reach an equitable solution, by connecting doctors' pay with other classes of people's pay (comparing), connecting increased pay with the cost of living rate, connecting pay with costs such as recruitment costs and other associated expenses, and connecting with consequences, such as the risk of underpaid doctors leaving the service or an inability to recruit experienced staff. We can also use ordering to rank the importance of these consequences versus the costs of pay increases. Connection thinking can similarly be used to allocate the government winter fuel allowance to elderly and disabled members of society, by considering whether there are any connecting rules to be adhered to (and whether to obey these or not), whether there are any connections regarding the amount of the payment given (such as cost of living, means testing, or whether to reduce, increase or maintain the payments), and whether there are any possible connecting mistakes which require precautions. You can then expand your set to include other types of government payments.

Over the following chapters, I will demonstrate how the global adoption of set thinking will fundamentally alter our culture and the way our society is governed, and how we can use it to develop workable, comprehensive solutions to the greatest threats facing the modern world, from climate change and wars to combatting the next pandemic. Society can be organised in a mathematical way in the way God would want rather than existing classes of governing as He is a mathematical being [p 71].

Politics and Governance

Adopting a mathematical approach using set thinking and the 15 classes of knowledge offers a viable alternative to the partisan debate politics that characterises modern governance.

Debates are not systematic. Their outcome is often heavily influenced by the policies of the political parties involved, the need to appeal to the interests of the electorate, the life experiences of the politicians, and their ability to persuade – their charm, likeability, fluency with words, and charisma. Debates are also frequently the cause of political conflict and result in ineffective leadership. Political ideologies create intransigence and dogged, inflexible policies. The Left takes one stance, and the Right another. They define themselves by their differences, refusing on principle to agree, and they hold to their beliefs of what will best serve the country in defiance of evidence to the contrary. Confirmation biases lead them to prioritise information that supports the beliefs they already hold and distrust sources that contradict them.

'Alternative facts' are nothing new and were certainly not invented by the Trump administration. We all present alternative facts; every time we argue for a political stance or a preferred way of doing things, we are selecting the facts that suit us and dismissing those that do not. In a society organised along mathematical principles, however, mathematical models and set thinking will equip politicians with a rigorous and efficient means of gathering all the relevant or connected knowledge on the problem at hand.

Let us consider an example. Two politicians are arguing about the state of healthcare in the UK. One argues that the NHS is in disarray, presenting as evidence a maternity hospital with poor performance. The other claims the NHS has never been better, highlighting an award-winning hospital as proof. Both these data sets are correct; it is the politicians' subjective interpretation of the data which is at fault. There exist both good and bad classes of hospitals. Each can be represented by a set of knowledge which can be gathered into sets, revealing the truth, and allowing us to dispense with the unhelpful rhetoric. With objective reasoning, we can dispense with the faulty logic that tells us there are two sides to every argument when the truth is that both sides may be correct for their given circumstances.

The circuitous and ideologically driven debates that typify our political system are a direct consequence of our politicians' inability to see the complete picture. Limited by their own knowledge and experience, they cherry-pick data to support their arguments, becoming divided along ideological lines into factions, each with their own fixed viewpoints. The fractious Brexit referendum is a prime example of this, with politicians often breaking from their own parties to join the Leave or Remain camps. Mathematical thinking will eradicate the drive towards such political manoeuvring. By applying set thinking, politicians can gather all knowledge for a subject, determining the policy objectives, ordering and prioritising policies, identifying mistakes and their precautions, factoring in all costs, value for money and other aspects of finance, identifying and connecting with all relevant types of people to be catered for (reality), and ensuring that appropriate rules are created and enforced, all consequences are identified, and all possible means of achieving an aim (actions) have been identified.

If our leaders rule mathematically, they can obtain a clear picture of reality on which to base their decisions. By working together, using mathematical models and set thinking to identify and implement successful policies, sharing mistakes and best practice, and using independent auditors to check their procedures, they can protect against mistakes, exaggerations, and the misappropriation of facts which result in ineffective or unnecessarily costly policies. With transparent processes, they can seek the input of third parties (universities, professionals, and experts) to gather all connected knowledge. They will ask themselves, is there sufficient evidence to prove that the information has truth characteristics? And with all the relevant information at hand, they can make informed choices and govern more wisely, and inclusively.

A recent television interview with the Secretary of State brought home to me the impact a lack of set thinking is having on the ability of our political institutions to address the myriad crises facing the world today. Asked about the government's climate change prevention policy, the Rt Hon Michael Gove described them as 'rigorous and informed by science'. Such a claim is intrinsically worrying to me. After all, without the benefit of an objective thinking strategy to accumulate all information on the subject and rationally evaluate it, how rigorous can it really be? Not to mention the rigorous application of mathematical principles will always supersede that of scientific enquiry. I am forced to ask, has every piece of information been thoroughly examined to ensure its validity, leaving nothing open to interpretation? Is the policy free from ideological leanings? Have checks and balances been put in place? Has a complete set of all the available options been considered and implemented? Considering that in my region of the UK only 5% of buildings have solar panels installed, I suspect not. Only by using set thinking and the 15 classes of knowledge can the solutions to climate change, or any issue, be fully considered, cost, and a comprehensive plan put in place.

By applying a mathematical approach to governing our societies, ideological differences, political parties, and biases become meaningless. Rather than taking a position and sticking to it, our leaders will gain the ability to consider each issue on its merits with full awareness of all the relevant facts. Instead of being entrenched in ideological certainties, they become open to a greater variety of options and can select the ones which prove most worthwhile. Rather than conflict, we can have agreement upon, and implementation of, best practice by sharing mathematical models, verifying truthfulness, and entrusting responsibility for best practice to a dependable person or organisation. And instead of a closeted decision-making process, where only the opinions of elected officials are considered, we can have a far more open and transparent system where informed outside parties such as universities, NGOs, and other specialists can make contributions. With their input, complete sets can be created which ensure all subsets of humans are included in the model to be connected with, mistakes can be predicted, and precautions taken, and information on best practice can be shared nationally and internationally to ensure everyone is singing from the same song sheet.

This task becomes easier still with the support of a less ideologically motivated electorate. Certain that our leaders are making reasoned, mathematically based decisions, we can focus on living our lives with no need to concern ourselves with issues beyond our immediate experience. Public opinion will no longer steer policies, as people will not hold opinions other than those obtained through direct experience, enabling our leaders to select policies based on their objectively assessed merits. And voting will only be necessary when there are crucial decisions to be made on policy directions, and on such occasions, transparent information sharing will enable us to make informed decisions. Imagine how different the Brexit referendum would have been had it been held in a mathematically organised UK. There would have been no factions, no strongly held opinions to divide the country, and certainly no abuse of facts to steer the debate. The outcome would have been determined by the facts, objectively sourced, tested, and shared, and not on perceptions. “What I have learnt is that there is an undeniable sense that politics just doesn’t work the way it should. The feeling that Westminster is a broken system.” So says the UK Prime Minister, Rishi Sunak. I have to say, I agree with him. Our subjectively organised system of governance has failed us. It is long past time to make a change.

As we become more mathematically literate, other ideologies will fall to the wayside, replaced by mathematical models, and set thinking. Democracy, communism, capitalism, religious and military leaderships, and dictatorships are all types of governments which could be replaced with a mathematical means of governance. This way of organising society will be more effective, and in line with what God would want for us. As a Russian priest said, ‘there is something ungodly about Western values’.

Our liberal democratic political model has its roots in the world’s first democratic experiment in ancient Athens, which was characterised by debate. Amphitheatres and the agora (town square) were used by the Athenians to host debates, take decisions, and participate in public life. Greek philosophy mirrors this emphasis on debate. Even Plato, who famously inscribed the words ‘Let no-one enter here who is ignorant of mathematics’ over the door of his academy, pursued knowledge entirely through discussion and argument (Magee, 2001). The contradiction of his approach is not lost on me. Arguments are not an effective means by which to determine reality. The only way to collect all knowledge for a given system or experience is through mathematical means, using the definition of a set, the 15 classes of knowledge and the mathematical properties of language, and the governance of our societies ought to reflect that.

A Mathematical Model for Governance

Let us now consider how politicians can make use of the 15 classes of knowledge to create workable policies, based on information that possesses truth characteristics.

Primary Classes of Knowledge

Reality

Defining reality is a crucial first step for politicians when developing policies. Consider all the relevant human subsets that ought to be connected with (catered for) by the policy, referring to probability to identify which subsets are the most likely to be impacted, and therefore most relevant. Depending on your policy, subsets could include anything from the visually impaired, the disabled, and those with learning difficulties, to people who own computers or do not. Also consider the animals, vegetation, fungi, inanimate objects, fire, atmosphere, gravity, and weather that ought to be connected with. Ask yourself, do you have a complete set?

Time

Consider when the optimal time to implement the policy will be. Are there other timing factors that ought to be taken into account? Does a timetable need to be created?

Action

Consider how the policy should be implemented. This can be established by conducting an audit. Ask yourself, do you have a complete set of all the available implementation methods?

Consequences

Consider the effect the policy will have. Will it impact some sections of the populace differently than others? Could it have adverse consequences? What will the benefits be? Will there be long-term impacts that differ from the immediate effects? Continue to list the possible consequences until you have a complete set.

Rules

Consider all the rules, regulations, and laws currently in existence, how these will shape your policy, and any challenges they may pose. You may also wish to introduce new rules or amend existing ones to make them more compatible. Ask yourself, are these enforceable, or will the free will of your electorate lead them to be ignored or circumvented?

Place

Consider where your policy will apply. Ask yourself, have you considered every community that it will impact? Or in other words, have you created a complete set?

Secondary Classes of Knowledge

Objective

Consider what your policy is intended to achieve. What will success look like? The experience can be represented by a set of knowledge and the objective will help to define the characteristics of that set.

Finance

Consider the costs of implementing your policy, any savings or financial efficiencies that will be gained by it, and any income it will generate. Include the costs of legislation changes, and secondary costs such as the losses of other forms of revenue. Ask yourself, is it a complete set? The methods of financing your policy also constitute a set. These may include reducing expenditure, increasing taxes, or grants and appeals.

Responsibility

Every aspect of the process should have a person or group assigned responsibility for it, ensuring nothing falls through the cracks. Consider who is the best person (or government department or organisation) to have responsibility for implementing and managing the policy overall, and for its constituent elements. Also consider what responsibilities other individuals or organisations will have. What is the government's responsibility, for instance? And what responsibilities will be allocated to voters, or taxpayers, or other subsets of society? Be certain to instruct the responsible person or organisation of their responsibilities. Ask yourself, have all the responsibilities been defined that ought to be? Or in mathematical terms, do you have a complete set? Has everyone been made aware of the responsibilities they now hold? They are given a list of their responsibilities.

Mistakes and Precautions

Conduct a risk assessment to establish which mistakes can occur and the precautions that can be taken, until you have a complete set. Then share your findings nationally and internationally.

Best Practice

Seek ideas from abroad or create best practice yourself by referring to trusted sources such as universities. Assign responsibility for sharing your best practice findings nationally and internationally to a reliable individual or group.

Information

Publicise relevant information about your policy so that the general public is aware of the changes, how it will impact them, and any new responsibilities they now have as a result.

Probability

Consider the likelihood of mistakes, such as natural disasters, occurring and take appropriate measures to prevent them where necessary. Also consider the likelihood of types of people (subsets) in the system (reality).

Truth

Conduct research or commission an audit to ensure the truth of the system, and whether you have considered every eventuality (created complete sets).

Quality

Ask yourself, is your policy planning now fit for purpose?

War and Conflict

Rejecting debate as an ineffective tool for decision-making enables us to resolve conflicts more efficiently and more peaceably.

Many of our most entrenched conflicts today are beset by long-held animosities that hinder attempts at resolution. Racial and religious hatred, the memories of past offences, and entrenched beliefs about their enemies can prevent both sides from forging their futures. Enmities between warring factions are understandable, but not helpful when seeking peace. In contrast to the current system of dispute resolution, which relies on mediated arguments and the subjective interpretation of historical and present-day facts, a popularity of ideas if you will, the 15 classes of knowledge offer a method for decision-making that is based on objectively tested facts. With both sides of the conflict living in the moment, unburdened by past experiences and historical grudges, they can work together to forge a path to a peaceful future. Without perceptions and generalisations to colour their interpretations of the 'enemy', they can view their opponents as a unique set of knowledge and see them as they really are

– human beings who are based on the same template as them, unique, and yet at the same time inherently the same. Equipped with the ability afforded by set thinking to judge each experience rationally and with an unbiased mindset, the warring factions can mediate for a peaceful solution to the conflict.

Set thinking offers a new approach to bring to the negotiating table, where facts are rationally examined, and the needs of both sides are assessed and attributed equal importance. Negotiating a fair and peaceful settlement is a complicated endeavour, but the task can be made immeasurably easier by utilising the 15 classes of knowledge to gather all connected knowledge about the cause of the conflict, the people involved, the needs and desires of all parties, and the consequences of proposed solutions. A complete set can then be created of all the possible solutions which are available to resolve the conflict, which can be assessed using all the mathematical tools at our disposal to select the best path to resolution.

Mathematical thinking can also act as a preventative measure for the onset of wars. Without stereotyped preconceptions and generalisations to warp our view of people who are different to us – in race, religion, or societal norms – the cause of many conflicts will be eradicated. Religious wars will also cease to occur, as the adoption of mathematical thinking will do away with the need for religious ideology and it is realised that God introduced religion as a stopgap before someone discovers mathematical thinking which is the way the mind should operate. Many conflicts today, and throughout history, have been either incited by the ideological differences between people of faith or justified by them. The most recent clashes between Israel and Palestine for instance have seen both sides claim that the conflict is the 'will of God'. By contrast, God's peaceful solution for all conflicts is for states to concentrate on governing their own people mathematically and helping other states through the sharing of mathematical models. Focused on living their lives moment by moment, unconcerned with irrelevant data, our leaders will be less inclined to meddle in the affairs and cultures of others. A recipe for lasting peace.

Culture

Perceptions are at the root of the prejudices and inequalities that beset our world. As our perceptions of reality are so often wrong, subjective thinking leads to misjudgements and mistakes. Racism, sexism, and gender biases all arise from a misrepresentation of reality based on faulty assumptions and incomplete facts, made by people who have attempted to understand the world and failed to realise that they do not.

Today's world is littered with accusations against social groups and people which are entirely unproven, yet the belief in them is often held so strongly, and by so many, that they can influence policies and behaviour on a massive scale. There exists all manner of offensive and damaging claims: Jewish people are stingy with money; people on benefits are lazy and do not want to work; skinheads are all racist; and women are worse drivers than men. If we try to remember where we first heard these generalisations, we will probably come up short. They are just things we have always known. But we must have heard them somewhere, and for them to be so ingrained, probably several times from several different sources. They are just part of the miasma of unsubstantiated data that surrounds us, seeping into our consciousness day after day.

An objective thinking strategy will free us from the burden of seeking reality and leave us less susceptible to the misjudgements and subjective representations that underpin such stereotypes, biases, and prejudices. By remaining content within the sphere of our own experiences, moment by moment, we can avoid this slow drip of inconsequential, yet highly toxic, information. By examining, testing, and evaluating information using rigorous mathematical techniques like set thinking and the 15 classes of knowledge, we guarantee that the information we base our beliefs and decisions on is truthful. Most generalisations would not stand up to this level of scrutiny.

Free from preconceptions, we can now view people as a unique set of knowledge and judge them on their own merits. As previously discussed, when you review the set of knowledge that represents a person and connect or rate each quality (ordering), you inevitably find that the most important qualities are those not directly discerned by the senses, such as a person's appearance, but rather their more hidden qualities such as loyalty and kindness (p53). You cannot always judge a book by its cover. Set thinking therefore challenges the preconceptions and inanities that underpin stereotypical thinking.

In discovering this, we learn something important about ourselves. That our own physical appearance, our body image, and how we are perceived by others is insignificant compared to who we are, and how we contribute to the world.

And with our minds uncluttered and free of perceptions, we are less biassed, less judgemental, and more comfortable in our own skins. With set thinking, we will become able to judge each experience rationally and with an unbiased mindset. Less concerned with the external pressures and expectations of the outside world, we will learn to be more accepting, less judgemental of others, and of ourselves, and to properly assess the consequences of our choices and actions. And with set creation opening our minds to new possibilities, our thinking processes become less rigid, and less constrained by the stereotypes we have been taught to believe in. We will retain our childlike flexibility and openness.

This change in the way we think will have an extraordinary impact on our culture, the way we treat and interact with others, and the way we view ourselves. Our self-esteem and our belief in our own sense of worth will be irrevocably altered. The impact will be greatest for the most vulnerable in our society, none more so than our children. Our children learn their concepts of beauty, desirability, and popularity through mirroring the evaluations we make, or that they see on TV and social media. This shapes how they feel about themselves and about other people. They denigrate themselves for not living up to the standards of perfection they are encouraged to set, and in turn, both resent other people who do and judge others who do not. The result is low self-esteem, bullying, and a fixation on celebrity culture. But it need not be this way.

Imagine a world where racist, sexist, and discriminatory beliefs were not a chronic stain upon our society; no longer institutionalised and endemic. Imagine a future where our children could accept themselves for who they are, conscious of their worth. No longer striving to live up to proscribed, idealised versions of perfection, but instead remaining focused on living their own lives, in their own way. This new culture of acceptance goes beyond tolerance of our differences, to an appreciation of them in ourselves and others. It is the basis for a healthier inner life for us and our children, and a more equal and just society.

Combating the Next Pandemic

When the Covid pandemic burst onto the scene with startling alacrity in the early months of 2020, it exposed how vulnerable the world continues to be to virus outbreaks. The need for a cohesive strategy for responding to such crises in the future is now an imperative. I would argue, however, that the crisis also exposed something else – the urgent need for objective thinking strategies to be adopted globally.

The crisis highlighted the fundamentally flawed nature of our current system of governance. Across the globe, parliamentary debates, and the associated political disputes, over the correct response to the pandemic may have slowed down, or hindered, the delivery of an effective response. In the UK, politicians debated the merits of taking the country into a lockdown, or not, how long these should last, and what the conditions of the lockdowns should be. Similar debates were held across the world. Few of these debates were grounded in a comprehensive review of all the facts, relying instead on fragmentary evidence or ideological priorities.

Debates continue to rage to this day into the effectiveness of the response, the steps that ought to be taken to protect against future outbreaks, and even into whether investigations should be held to review the response and learn lessons for the future. In the US, a group of Senators recently tried to revive efforts for a national Covid-19 commission along the lines of the 9/11 investigation, only for their attempts to be marred by partisan arguments over the cause of the outbreak and the effectiveness of distancing measures.

The emergence of Covid sceptics, whose membership included some governments, and whose opinions were in opposition to medical and scientific evidence, is also a clear example of the potential impact of subjective thinking. While the vastly different strategies imposed across the globe highlight the failure of States to effectively share information and best practice.

Effectively combating future pandemics will require the application of set thinking and the 15 classes of knowledge. These objective thinking strategies are the only means by which we can gather all the connected knowledge required to create systematic plans which learn from past mistakes and best practice, and protect the entirety of society equally and inclusively.

A Mathematical Model for Combating the Next Pandemic

I will now take you through the step-by-step process for devising a coherent response to an emerging pandemic using mathematical means. By now, you should be familiar with the processes involved. You can use this case study as an opportunity to hone your own newly acquired set thinking skills. Why not complete the mathematical model as we go, creating your own complete sets to add to the details provided here.

Primary Classes of Knowledge

Reality

Consider who you should connect to. Human subjects will include doctors, nurses, and patients, but you can also take this further by connecting to all the subsets of humans affected, such as the disabled, the elderly, those with pre-existing medical conditions, school-age children, teenagers, parents, care-workers, and employers. Consider all the myriad demographics present in your subject sample, such as professions (those in the food and transport industries for instance), race, gender, and age. Also, consider any groups of people that may be relevant, such as prison inmates, attendees at weddings, funerals, restaurants and other gatherings, hospitals, care-homes, public houses, or schools. These may be relevant either as disease spreaders or as vulnerable populations. Ask yourself, do you have a complete list of all groups of people?

Now consider your non-human subjects – animals, vegetation, fungi, fire, atmosphere, weather, gravity, and inanimate objects. The latter will be especially important when developing a pandemic response, as it encompasses all the equipment that will be needed, such as personal protective equipment (PPE), vaccinations, medicine, and hospital beds, as well as any inanimate objects which may be impacted by the outbreak, such as products and supplies whose manufacture or transport may be affected. The weather may also be a contributing factor to the spread of the virus.

Time

Consider all matters relating to the timing of your virus response. Should certain elements of your response be put in place in advance of an outbreak? You may decide, for instance, to stockpile equipment such as medical supplies, vaccinations, and PPEs. Also, consider the timings around the roll-out of your response, taking into account which aspects should be rolled out first, which will take longer to implement, and any time-sensitive actions. You may want to create a timetable, or schedule, of response activities to be put into effect in the event of an outbreak. The season may also affect access to resources, such as the greater demand on hospital beds during the winter months. Ensure you have a complete set before proceeding.

Action

Consider how your response will be implemented. This could include testing people for the virus, acquiring PPE, training health staff, increasing hospital capacity, and running vaccination programs. Ask yourself, have you considered every possible action you, or others, could take to improve your response to the outbreak? Or in other words, do you have a complete set?

Consequences

Now carefully consider the consequences of these actions. Will your vaccination program draw healthcare workers away from other vital tasks? Will your plan for the infected pose a contamination risk to other patients, medical staff, or members of the public? As always, use your set thinking tools to classify and expand upon every variable and remember to ask yourself if you have a complete set.

Let us consider an example. You identify that a possible consequence of moving elderly patients from hospitals into care homes – a daily occurrence under normal conditions in the NHS and one that is essential to ensure beds are available for new patients – is the risk they may spread Covid to vulnerable residents. You, therefore, consider all precautions that could be taken to avoid this, such as mandatory Covid testing before discharging patients. You consider the possible consequences of this preventive measure (costs, delays, wasted resources) and use ordering to determine whether the risk justifies them and to select the most efficient response.

You can then raise the classification, applying classify and expand techniques to consider which other group situations people may enter where Covid testing could be used to prevent spreading the virus. You may wish to consider psychological consequences as well as the purely physical or practical ones. Will your quarantine procedures impact the populations' mental health, for instance, or lead to emotional distress by preventing people from visiting dying loved ones or forcing women to give birth without familial support?

Rules

Consider all the rules, regulations, rules of thumb, and laws that are relevant to your response. These may include British Medical Association guidelines, employment laws, and if you are importing equipment from abroad, trade regulations. Ask yourself, do you have a complete set? Now consider whether (or not) these rules should be obeyed, and what the consequences will be of not doing so, and consider whether new rules (with the characteristics of being enforceable) will need to be introduced.

Place

Consider all the connected locations. These will include where illnesses may occur, where they will be treated and prevented, where groups of people who are most susceptible to the virus or at the highest risk of becoming virus spreaders will be located (such as care homes and prisons), immunisation and testing sites, and other locations for your outbreak response. This may include hospitals, doctor's surgeries, and repurposed locations such as shopping centres and town halls. You may also want to consider the location of your base of operations from where the response will be directed. Once you have a complete set, you can move on to the secondary classes of knowledge.

Secondary Classes of Knowledge

Objective

Your objective here is clear: to create an effective plan for a future pandemic. You may also want to define what you mean by an effective plan by listing key performance indicators for success that you will work towards achieving.

Finance

Consider the costs of implementing your response until you are satisfied you have a complete set. These should include direct costs such as the purchase of PPE, vaccinations, research, and administration, and indirect costs such as the impact on the wider economy.

Responsibility

Consider where responsibility lies for the implementation of your response. Will implementation require parliamentary consensus, for instance, or can it be triggered by a decree from the Prime Minister or another leader? Who will be responsible for managing the response? A government department perhaps, or a specially appointed tsar?

You should also consider what other responsibilities will need to be allocated and to whom. You may wish to divide tasks between different government departments, research organisations, and bodies, or retain control within a centralised body. Now ask yourself, do you have a complete set of all the posts that require filling? Is everyone involved aware of the responsibilities they have been assigned?

Mistakes and Precautions

Consider the mistakes which may occur. These may include the purchasing of PPE which is unsuitable for the purpose, misallocation of funds and over-spending, a failure to make adequate provisions for some of your subject subsets (such as people in nursing homes), or failures in your testing regimens. Ask yourself, do you have a complete set of mistakes and precautions?

Use set thinking to determine which failures have the greatest probability of occurring, with the worst consequences, and factor this into your prevention planning. The numerous investigations into the Covid-19 response which have been conducted around the world could be a useful source of information gathering here. Refer to these and other learnings from Covid to help you create a complete set, then share your results nationally and internationally.

Best Practice

The Covid-19 response could also provide useful information on best practice, as can the responses to other outbreaks around the world. Seek as much evidence as possible, from multiple sources. Once you are confident you have a complete set, share your findings nationally and internationally.

Information

Sharing information about your pandemic response will be critical to its smooth implementation. Ensure that everyone who is allocated responsibilities by the plan is aware of the roles they will be expected to play, and how to do so. This may include the general public, who will benefit from preparedness and be reassured by the knowledge that a plan has been put in place.

Probability

Consider the likelihood of a pandemic occurring, of each of the mistakes you identified being made, and of factors that may prevent the swift implementation of your plan. If necessary, revise your planning in light of your discoveries.

Truth

Make use of auditors or quality assurance officers to establish the truthfulness of the system and ensure your plans are workable. This may include research into the suitability of your chosen PPE equipment, to establish which controls ought to be put in place to prevent mistakes, or into the feasibility of increasing hospital capacity to meet your needs.

Quality

Ask yourself, have you produced a satisfactory means of gathering all knowledge needed to prepare for a future pandemic? If the answer is yes, then you should now be in possession of a comprehensive and effective plan for combating the next pandemic to strike our world.

Climate Change

Climate change is one of the greatest threats to our modern world. There is now more CO2 in our atmosphere than at any point in human history – levels unseen on Earth for four million years. The latest scientific results predict that if global temperatures rise just 1.5 degrees above pre-industrial levels, it will unleash a cascade of increasingly catastrophic, and potentially irreversible, impacts – and we are set to exceed that temperature within the next five years. Yet, the global response to this immense challenge has been uncoordinated and grossly disappointing. Numerous attempts to create a unified international response to the crisis (such as the 1992 UN Framework on Climate Change and the 2015 Paris Agreement) have generated empty promises that few countries have made significant strides towards implementing. Climate reduction efforts in the UK have been similarly lax. A recent Climate Change Committee report described government efforts to scale up climate action as 'worryingly slow'. This despite the 2022 climate change risk assessment warning that the UK faces eight immediate high risks due to climate change, including risks to soil health from increased flooding and drought; risks to food supplies, goods, and vital services due to supply chain collapse; risks to people and the economy due to the failure of our power systems; and risks to human health, wellbeing, and productivity from increased exposure to heat in homes and other buildings.

A mathematical approach must be adopted as a matter of urgency, by our own government and those around the world, if we are to find a workable solution to the crisis we are now facing. We need a comprehensive global strategy for tackling climate change, with the commitment of the entire global community. Set thinking and the 15 classes of knowledge offer us the tools by which this can be achieved.

A Mathematical Model for Tackling Climate Change

You can apply set thinking to classify all the actions that the world's governments could take to reduce climate change, expanding out to include installing wind turbines, harnessing tidal power, installing solar panels, growing seagrass, issuing grants for climate reduction measures, carbon dioxide capture and a myriad of other available options, until you have a complete set.

You may want to connect with members of the scientific community, climate change experts, and the business community to help identify all the methods at your disposal. You can also use ordering to determine the best climate prevention measures, evaluate the consequences of each option, and adopt permutation and combination to create a tailor-made solution that utilises several climate change prevention measures. You could create a downward set for each of these options, i.e., panels of solar panels etc. You may also use upward sets such as places where solar panels can be installed, restricting this list in any way that suits your needs, (for instance, a set of buildings, or fields, where solar panels could be installed with X characteristics). The same process could be used to define the characteristics of a set where seagrass should be planted, identifying the best ones by classifying and expanding. Once you have classified and expanded to create a set of actions the government could take to reduce climate impact, you can use ordering to determine which of your options are the most viable.

You could also classify a matter that the general public could do to support the reduction in CO2, expanding to include installing an eco-boiler, insulating their homes, driving an electric car, switching to public transport, eating less meat, and using car-sharing.

To ensure you have all the information necessary to devise your global climate change strategy, you will then need to apply the 15 classes of knowledge to each one of these solutions in turn. Let us take installing solar panels as an example. Identify complete sets for *Reality* (including solar panels, the weather, and homeowners); *Time* (when to install and operate the solar panels); *Action* (installing solar panels, publicising the policy and issuing homeowner grants); *Consequences* (generating green power, reducing CO2 emissions, public concerns about the project); *Rules* (planning regulations, manufacturer's instructions, and the impact and consequences of users' free-will decisions); *Place* (where to install the solar panels); *Objective* (reducing CO2 emissions); *Finance* (the cost of the solar panels, their installation, and maintenance, the impact on property prices, and the opportunity to apply for grants);

Responsibility (assigning a suitable team or individual to oversee the project and defining homeowner responsibilities);

Mistakes and precautions (poor quality panels, or lack of value for money, scams etc); *Best Practice* (seeking best practice from abroad e.g., solar panels that operate with light as well as sun, or creating your own list); *Information* (seek relevant or connected knowledge from the internet and trusted sources); *Probability* (the likelihood of mistakes, and other factors that may influence project success such as natural disasters); *Truth* (conduct research to ascertain the truth of the role solar panels can play in preventing climate change); *Quality* (Ask yourself, have you identified a satisfactory way of gathering all knowledge for this means of climate change prevention?). Having completed your assessment of installing solar panels (gathering all relevant and connected knowledge), you are ready to circulate your mathematical model nationally and internationally, making certain to share your best practice findings, mistakes, and precautions.

With this task complete, you can now move on to evaluating the other actions that governments can take to reduce climate change which you identified using classify and expand set thinking tools. You can then begin ordering them and utilising permutations and combinations. Eventually, you will have at your disposal all the information you will need to develop a comprehensive strategy to combat climate change, once and for all.

Conclusion

The widespread adoption of objective thinking strategies would be a watershed moment in human history, changing the direction of our society forever. It is a reset switch. An opportunity to discard those aspects of our culture which have proven unfit for purpose and start afresh. A chance to live life to the fullest, unburdened by the clutter of irrelevant and useless perceptions which distract us from living our lives.

Outside my local church in Wallasey, a sign welcomes the congregation with the phrase 'God's free gift is a new start'. It is a poignant reminder to me of the positive changes we can bring to this world if we begin using our brains with the mathematical precision He always intended. God has gifted us with the mathematical tools we need to remake our world, and a mind capable of cogitating with mathematical precision. It is beyond time that we began making full use of those gifts.

Changing how we think will ultimately result in a need to restructure our society, raising new questions about our political systems and the way our public life is organised. This is nothing to be afraid of. Society is constantly adapting. Lifestyles change, views alter, technology advances, and society shifts with them. With our newfound objective reasoning, we can make this transition smoothly, and without the need for arguments or hostile debates. Our new society will emerge organically, changed from within to best represent the people who call it home. A magical mathematical world, just as God always intended for us.

Bibliography

Chalmers, D. (2022), "Virtual Reality Is as Real as Physical Reality but Just Different," New Scientist, 29 January, pp. 14.

Chalmers, D. & Koch, C. (2023), "A 25-Year Bet about Consciousness Has Finally Been Settled," Scientific American.

Comte-Sponville, A. (2004), The Little Book of Philosophy. London: William Heinemann.

Magee, B. (2001), The Story of Philosophy. London: Dorling Kindersley.

Nagel, T. (1986), The View from Nowhere. Oxford: Oxford University Press.

Penrose, R. (2005), The Road to Reality: A Complete Guide to the Laws of the Universe. Surrey: Vantage.

Penrose, R. (2006), "What Is Reality," New Scientist, 18 November, pp. 5.

Spiegel, M. (1963), Theory and Problems of Advanced Calculus. Wisconsin: Schaum.

Thompson, M. (2003), Teach Yourself Philosophy. Oxfordshire: Bookpoint.

In our hour of need, as climate change, conflict, and economic turmoil threaten our very survival and our mental health crumbles under the strain of not using our minds in the way God intended, Francis Keith Robins offers a revolutionary new approach to saving humanity – transforming the way we think through mathematical means.

In *The Fourth Coming*, Robins presents an intimate step-by-step guide to unlocking our mathematical natures to create the inclusive, equal, and peaceful future God wants for us. One where shared mathematical models and set thinking replace the failed classes of governments and institutions that threaten our world, changing the course of human history in the way God would want. A work of mathematical genius, his theories on God mathematics posit a radical, simple solution to consciousness – per New Scientist, one of the last remaining and greatest mysteries to science and philosophy. He also concludes that God importantly introduced religion because people were not using their brains in the way He intended.